# Rhenium Disulfide

**David J. Fisher**

Published by **Materials Research Forum LLC**
Millersville, PA 17551, USA

Published as part of the book series
**Materials Research Foundations**
Volume 40 (2018)
ISSN 2471-8890 (Print)
ISSN 2471-8904 (Online)

Print ISBN 978-1-945291-92-0
ePDF ISBN 978-1-945291-93-7

Distributed worldwide by

**Materials Research Forum LLC**
105 Springdale Lane
Millersville, PA 17551
USA
http://www.mrforum.com

Printed in the United States of America
10 9 8 7 6 5 4 3 2 1

# Table of Contents

Rhenium Disulfide                                    Materials Research Forum LLC
Materials Research Foundations **40** (2018)          doi: http://dx.doi.org/10.21741/9781945291920

# Introduction

Two of the major trends in materials science during the past decade have been the development of exciting new properties by manufacturing traditional materials in novel forms (e.g. carbon as graphene, etc.) or by the identification of useful properties in apparently mundane substances (e.g. $MgB_2$) or in substances better known for their other surprising properties (e.g. semiconduction in Heusler alloys). Rhenium disulfide is a prime example of at least two of those trends. It is now the subject of lively research into its electronic properties: especially when in a low-dimensional form. The field of two-dimensional materials such as graphene and its analogues has been growing very rapidly during the past fifteen years, with continuous innovations being made in new materials and novel devices. Most of these materials have quite symmetrical two-dimensional crystal structures, thus imparting similar electrical and optical properties to the various in-plane crystalline directions. Also appearing has been the class of two-dimensional layered materials. These possess low-symmetry crystal lattices, and include black phosphorus and its arsenic alloys. This class also includes rhenium disulfide and rhenium diselenide, which belong to the transition-metal dichalcogenide family. Due to their reduced crystal symmetry, they exhibit distinct electrical and optical characteristics along certain in-plane crystal directions.

Transition-metal chalcogenides have been studied since 1960, although rhenium disulfide was hardly mentioned at that time, when they then already amounted to 40 in number[1]. The group-VI members such as molybdenum and tungsten are the most typical ones, but group-VII rhenium disulfide has been attracting most attention of late because of its unusual structural, electro-optical and chemical properties; especially an indirect-to-direct band-gap transition which occurs when thinned down from bulk to monolayer. The group-VI transition-metal dichalcogenides have a 1H, 2H, 3R or 1T structure, whereas $ReS_2$ has a distorted 1T structure which imparts an in-plane anisotropy to its physical properties. Few other materials (black phosphorus, $ReSe_2$, $TiS_3$, $ZrS_3$) exhibit such an in-plane structural anisotropy. This makes $ReS_2$ unique among the transition-metal chalcogenides. Atomically thin rhenium disulphide is characterized by weak interlayer coupling and a distorted 1T structure, which leads to the anisotropy in optical and electrical properties. It also possesses structural and vibrational anisotropy, layer-independent electrical and optical properties and metal-free magnetism. In these respects, it differs from group-VI transition-metal dichalcogenides such as $MoS_2$, $MoSe_2$, $WS_2$ and $WSe_2$. It is already being used in solid-state electronics, catalysis, energy storage and energy-harvesting applications.

Materials Research Forum LLC
doi: http://dx.doi.org/10.21741/9781945291920

Monolayer transition-metal chalcogenides are two-dimensional materials consisting of an atomic plane of a transition metal (molybdenum, tungsten, titanium, niobium, rhenium, vanadium, zirconium, tantalum, hafnium, etc.) sandwiched between two chalcogen (sulfur, selenium, tellurium) planes A wide variety of transition-metal chalcogenides can therefore be obtained which possess diverse controllable electronic properties. The band-structure engineering of two-dimensional metal dichalcogenides is crucial to their light-matter interaction in opto-electronic applications. The addition of various metallic or chalcogen elements provides a versatile and efficient means for modulating the electronic structures and properties of two-dimensional materials. In such alloys, the quantification of spatial distributions and of the local coordination of atoms aids the understanding structure-property relationships at the atomic scale. Rhenium disulfide also attracts attention because of its extremely weak interlayer coupling. As a new two-dimensional semiconductor, rhenium disulfide offers many distinctive features and possible future novel device creation due to its unusual structure and unique anisotropic properties. A review of the recent progress in rhenium disulfide research is therefore in order.

**Chemical Synthesis**

In the very earliest relevant studies, the heat of combustion of rhenium disulfide (table 1) was measured by means of bomb calorimetry[2]. The heat and free energy of formation at 25C were found to be -42.7 and -41.5kcal/mol, respectively. The vapor pressures are given by table 2[3].

*Table 1 Energy of combustion of ReS₂ at 25C*

| Sample Weight (mg) | Heat (cal) | $\Delta$E (cal/g) |
|---|---|---|
| 64.89 | 65.984 | 1.0169 |
| 55.73 | 57.089 | 1.0244 |
| 61.82 | 62.463 | 1.0104 |
| 60.35 | 61.336 | 1.0162 |

*Table 2 High-temperature vapour pressures of ReS$_2$*

| Temperature (K) | Pressure (mm) |
|:---:|:---:|
| 1383 | 13 |
| 1462 | 55 |
| 1498 | 96 |

The homogeneity range of rhenium disulfide at 873 to 1473K was determined by vapor-pressure measurements using the dew-point method[4]. The rhenium disulfide composition with respect to sulfur, within the limits of the homogeneity region, was found to vary from 25.3 to 29wt%. At 873 to 1473K, the sulfur vapor pressure over alloys at the boundary of the homogeneity region on the sulfur side amounted to between 5 x 10$^5$ and 25.3 x 10$^5$Pa. At the boundary on the metal side, it decreased to 10$^{-1}$ to 5 x 10$^2$Pa. Many transition-metal chalcogenides are difficult to produce because of the high melting-points of the metals and of the oxide precursors. Molten-salt methods have been used to produce ceramic powders at quite low temperatures. Molten-salt assisted chemical vapour deposition can be used for the synthesis of a wide range of two-dimensional atomically-thin transition-metal chalcogenides. Binary compounds, based upon rhenium and 31 other metals, have been prepared[5]. The salt decreased the melting-points of the reactants and aided the formation of intermediate products; thus increasing the overall reaction rate. Most of the materials which were synthesized were useful, as exemplified by a high mobility in ReS$_2$. An early method for the synthesis of rhenium disulfide was the reaction of potassium perrhenate (NH$_4$ReO$_4$) with sulfur in a melt of alkali metal sulfides[6]. It was concluded that the ReS$_2$ which was obtained by using this method, or one involving the reaction between metallic rhenium and sulfur or hydrogen sulfide, crystallized in the hexagonal system. Direct sulfidation of ammonium perrhenate can lead to the synthesis of rhenium disulfide[7]. This is the first example of a simple bottom-up approach to the chemical synthesis of crystalline ReS$_2$. The reaction occurs at room temperature in a solvent-free environment, without requiring a catalyst, and leads to the formation of lower-symmetry (1T′) ReS$_2$ with a low degree of layer stacking.

The sulfidation of thin films of the transition metals, molybdenum, tungsten, rhenium, niobium and tantalum, has been used to study the synthesis of transition-metal dichalcogenides[8]. Metal layers were prepared by direct-current magnetron sputtering and were sulfidized by using sulfur vapor and H$_2$S. Photo-electron spectroscopy revealed that, following treatment with sulfur vapor, with p$_{S2}$ = 1 to 10Torr, the molybdenum, tungsten and rhenium films were transformed into $MoS_2$, $WS_2$ and $ReS_2$, respectively. Raman spectroscopy revealed an improved crystallinity in the case of molybdenum, tungsten and rhenium which was sulfidized in $H_2S$. Isobaric and isothermal stability plots identified suitable domains of sulfur partial pressure and temperature for the sulfidation of the metals (figure 1). In the case of molybdenum, tungsten and rhenium, an $S_2$ partial pressure of $10^{-5}$bar was sufficient to convert the metals into sulfide phases at 750C.

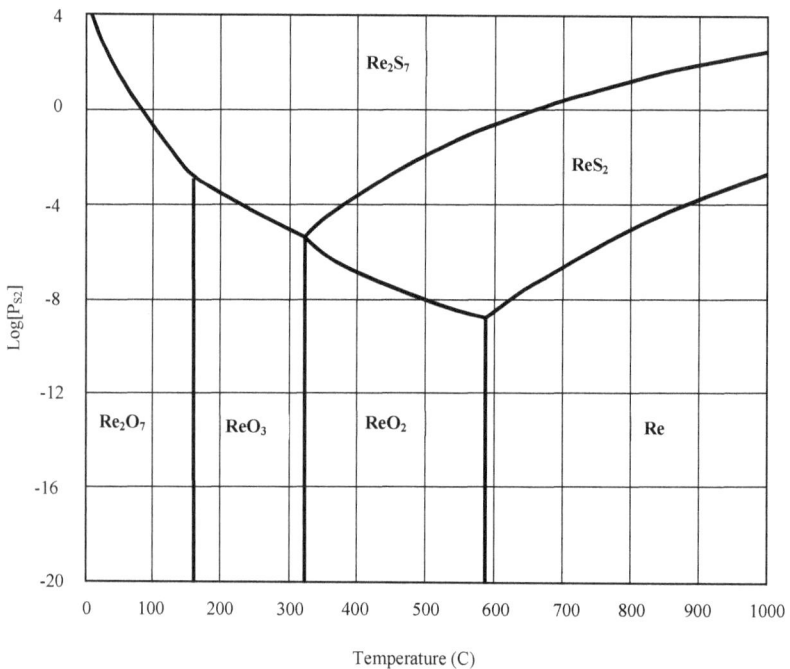

*Figure 1 Calculated isobaric stability diagram for the Re–S–O system*

Rhenium sulfide of poor crystalline quality has been prepared[9] via the solvothermal oxidative decarbonylation of the respective metal carbonyl by using sulfur and a p-xylene solvent. The same reaction, using hexadecylamine, was used to prepare a $ReS_2$/hexadecylamine nanocomposite. X-ray diffraction analysis, scanning electron microscopy, atomic force microscopy and Fourier-transform infra-red spectroscopy showed that $MoS_2$ and $ReS_2$ were structurally similar but morphologically different: consisting of undefined 150 to 300nm particles and well-defined almost perfect 0.4 to 2.8μm microspheres, respectively. Preparations which contained hexadecylamine produced corresponding layered materials. The $MoS_2$/hexadecylamine nanocomposite was a dark solid which could be easily separated from the reaction mixture. The $ReS_2$/hexadecylamine combination remained as a suspension in p-xylene, but could be separated by evaporating the solvent under vacuum. Both materials were layered, with a basal spacing of 33.8Å in the case of molybdenum and 30.4Å in the case of rhenium. Preparation of thin-film $ReS_2$/hexadecylamine from its suspension, by evaporation of the solvent in air, produced 0.4μm x 1.0μm cylindrical particles. A proposed process for the growth of rhenium disulfide monocrystals[10] involves mixing rhenium trisulfide and excess powdered sulfur in a suitable ratio and placing them in a quartz tube under vacuum and maintaining them at 400C for up to 48h. The resultant powder is then ground, and burned on an alcohol burner to remove any remaining sulfur, thus leaving high-purity polycrystalline rhenium disulfide. The latter is then mixed with bromine as a transport agent before being sealed in a quartz tube under vacuum and heated in a multi-temperature furnace; thus growing rhenium disulfide monocrystal via the chemical gaseous transport method.

**Sample Preparation**

Rhenium disulfide has been attracting attention due to the direct band-gap which exists, regardless of its thickness, and due to the anisotropic electrical, mechanical and optical properties which arise from its unique crystal lattice structure. Some preparation methods are however not suitable for practical applications because of non-uniformity, discontinuity and the general difficulty of large-area continuous film growth. Before looking at the details of the preparation methods, it will be useful to offer a general summary. The bulk material consists of a stack of decoupled monolayers, so that both top-down and bottom-up fabrication of two-dimensional nanostructures are feasible. The common means of preparation of $ReS_2$ nanostructures are mechanical exfoliation, chemical vapor deposition and chemical or liquid exfoliation. One method involves halogen-assisted vapor transport at about 1000C, followed by the top-down use of mechanical or chemical exfoliation. Another method involves the bottom-up formation of monolayers via chemical vapor deposition. An established technique for preparing two-dimensional nanosheets is to use adhesive tape to peel off nanoflakes from bulk disulfide surfaces. Such mechanical exfoliation produces the highest-quality monolayers and can be used to fabricate high-performance devices. The so-called sticky-tape technique makes it difficult however to control the layer thickness and uniformity or to carry out large-scale nanosheet production.

These problems are avoided by using chemical or liquid exfoliation. In the former technique, ions of a selected metal such as lithium are intercalated between the layers in order to promote exfoliation. This is a good solvent-free method for producing highly controlled $ReS_2$ nanosheets. Lithium intercalation unfortunately leads to negatively charged nanosheets rather than the neutral ones which are desired. The problems of lithium intercalation and post-exfoliation modification have been partially addressed by growing thin films of the disulfide via chemical vapor deposition, followed by exfoliation of those films in N-methyl-2-pyrrolidone under ultrasonification. In chemical vapor deposition, the solid precursors are heated to produce vapors which are then reacted to form large-area crystals on a substrate. This offers scalability and reproducibility, but high-temperature treatment – imposed by the 3180C melting point of rhenium - may have to be maintained for weeks; all the time ensuring a flow of bromine or iodine. The high melting point of rhenium also decreases its vapor pressure and thus markedly reduces the product yield. Chemical exfoliation involves moreover a phase transition, which means that heat treatment is required in order to recover the original phase.

Liquid-phase exfoliation is a technique in which the bulk material is dissolved in a suitable solvent and sonificated for hours. This technique is not of much use for $ReS_2$ because of the lack of a solvent possessing a similar surface tension, and of the difficulty of controlling the product thickness. Some success has been had in preparing nanomaterials by means of liquid exfoliation, and layer-by-layer sorting in aqueous surfactant solutions, using so-called isopynic density-gradient ultra-centrifugation.

In order to avoid the problems caused by the high melting point and low vapor pressure of rhenium powder, ammonium perrhenate ($NH_4ReO_4$) is used as a source. The +7 valence state of rhenium and the appearance of unwanted by-products are the main cause of a poor crystal quality of perrhenate-derived $ReS_2$. Halogen-vapor transport leads to unintentional doping and thus affects the resultant electrical properties, with iodine giving p-type doping and bromine tending to give n-type doping. Chemical vapor deposition will however outstrip all other preparation methods for two-dimensional nanostructures if high temperatures and halogen-vapor transport problems were to be eliminated.

The temperatures involved can be reduced to 430C by using an alloy, such as a rhenium-tellurium eutectic, having a lower melting point; leading to large-area high-crystallinity uniform-thickness monolayers. Large-area highly-crystalline material can be produced via vapor-vapor reaction between sulfur and rhenium at 460 to 900C, using a mica substrate to grow uniform monolayers. Large-area epitaxial growth on a mica support at

Materials Research Forum LLC
doi: http://dx.doi.org/10.21741/9781945291920

500 to 800C can be achieved by using $ReO_3$ as a rhenium precursor. A drawback here can be the use of extremely operator-harmful hydrofluoric acid as an etchant to separate the diselenide film from the mica substrate. In order to avoid the use of halogen vapor, recourse can be had to the Bridgman method, which requires no transport agent but may need weeks to be spent at 900 to 1100C.

The preparation of various nano-forms of $ReS_2$ has been investigated. It is found that there are two growth directions. Growth in the (100)-direction is the predominant one, and is rapid. Growth in the (020)-direction is relatively slow. This growth asymmetry favors the appearance of nanorod-like one-dimensional products. In order to form instead a sheet-like hexagonal structure, the growth rate in both directions should be comparable. This can fortunately be arranged by varying the ambient concentrations of nitrogen and hydrogen. Thus a ratio of 10:1 produces nanorods while a ratio of 2:1 produces hexagons. Greater hydrogen concentrations decrease the planar density by introducing sulfur vacancies, increase the number of unpaired bonds and increase the energy of the plane; thus reducing its growth rate. At a critical concentration, the growth rate of the (100) plane then becomes comparable to that of the (020) plane and hexagonal flakes are the result. For the purposes of self-assembly and edge-to-edge assembly, the interlayer coupling of two-dimensional materials should preferably be weak. The layers of $ReS_2$ are coupled by extremely weak van der Waals forces which permit two neighboring nanoflakes to slide with no friction. By taking advantage of the anisotropic and weak interlayer coupling, self-assembled nanoflakes and nano-scrolls have been prepared by using electrochemical lithium intercalation. Four factors are involved: adjacent nanoflakes stack face-to-face, randomly-distributed nanoflakes retain their disordered structure, pre-existing nanoflakes tend to slide away under the influence of combined anisotropic and electrostatic forces and nanoflakes having coherent facets approach each other and fuse to form nano-scrolls.

Rhenium disulfide is a direct-gap diamagnetic semiconductor and its valence and conduction band edges are composed of the d-orbitals of rhenium atoms and the p-orbitals of sulfur atoms. The band structure of mono-, tri- and penta- layer material, calculated using *ab initio* methods, shows that the overall band-gap does not change significantly between monolayer and multilayer. This offers the interesting possibility of producing relatively large-area monolayers; both monolayer and bulk $ReS_2$ are direct-gap semiconductors with band-gaps of 1.55 and 1.47eV, respectively. The overlapping of electron wave-functions from adjacent layers is so weak that modulation of the interlayer distance cannot renormalize the band structure, implying that the material is electronically decoupled. Other transition-metal chalcogenides exhibit a crossover from an indirect to a direct band-gap in going from bulk to monolayer. The thinning of $ReS_2$

does not change its bulk-phase direct band-gap and, while other transition-metal chalcogenides suffer out-of-plane quantum confinement with decreasing thickness, the quantum confinement of $ReS_2$ is essentially independent of the number of layers. This is because neighboring monolayers in the flake are largely decoupled electronically, so that thinning the flake does not increase the quantum confinement of electrons.

The room-temperature Hall mobility of n-type material and the impurity carrier activation energy are reported to be $19 cm^2/V$ and 178meV. The intrinsic carrier mobility and resistivity depends upon the temperature and the electron density. The carrier mobility is inversely related to the temperature. The present material therefore exhibits a semiconducting-to-metallic behavior at high electron densities and low temperatures since, at high electron densities, the resistivity greatly decreases. The metallic state of $ReS_2$ is a result of a second-order metal-to-insulator transition driven by electronic correlation; illustrating the susceptibility of the band structure to an applied electric field. The band-gap and electronic transport can also be modulated by applying strain to the various axes of the unit cell. Thus tensile straining lowers the conduction-band minimum and raises the valence-band maximum, giving a reduced band-gap. Straining can greatly change the effective mass and the hole mobility while having a relatively small effect upon the effective mass and electron mobility. Here, a-axis straining increases the hole effective mass but decreases the hole mobility, while b-axis straining decreases the hole effective mass but increases hole mobility.

The low crystal symmetry of this van der Waals compound leads to highly anisotropic optical, vibrational and transport behaviors. A momentum-resolved study of the electronic structures of the monolayer, bilayer and bulk rhenium disulfide, using k-space photo-emission microscopy and first-principles calculations demonstrated[11] that the valence electrons in the bulk disulfide are appreciably delocalized across the van der Waals gap. Direct observation of the evolution of the valence-band dispersion, as a function of the number of layers, revealed a transition from an indirect band-gap in the bulk disulfide to a direct gap in the bilayers and monolayers. There was also an appreciably increased effective hole mass in monolayer crystals.

The reduced crystal symmetry of $ReS_2$ gives rise to anisotropic in-plane optical properties, and its triclinic structure leads to polarization-dependent optical absorption. The anisotropic absorption coefficient and transient absorption are greatest when light polarization is parallel to the rhenium chains and are smallest when it is perpendicular. Many-body perturbation theory shows that the lowest-energy bright excitons of the distorted 1T material exhibit a perfect figure-of-eight polarization. This is unusual among hexagonal transition-metal chalcogenides, and $ReS_2$ is active in capturing photons in the near-infrared regime. As well as absorption anisotropy, the photoluminescence emission spectra are very different to those of conventional transition-metal chalcogenides. In group-VI transition-metal chalcogenides, the emission intensity is an order of magnitude greater for monolayers - as compared with the bulk - due to a crossover from an indirect to a direct band-gap. The intensity increases with the number of layers of $ReS_2$. In the optical absorption spectra, 3 optical transitions are identified in both bulk and monolayer material, indicating a decoupling of the excitonic and emission properties.

The introduction of foreign metal or non-metal atoms as dopants can be achieved by replacing sulfur or rhenium atoms. Among non-metals, chlorine is reported to be the best candidate for n-type doping while molybdenum and phosphorus are the best candidates for p-type doping monolayers. Doping can increase the electrical conductivity by increasing the carrier density. Fluorine can be used to tailor the electronic and magnetic properties. The presence of fluorine atoms above the rhenium chains can induce ferromagnetically-coupled metallic mid-gap states within the rhenium chains while antiferromagnetically coupling the rhenium chains. On the other hand, fluorine atoms between the rhenium chains generated semiconducting mid-gap states and were non-magnetic. Fluorine-treated material is therefore promising for spintronic and spin-wave logic applications. The net magnetic moment of doped material ranges from 0 to $1\mu_B$. A red-shift in the optical absorption is observed after niobium-doping but the direct band-edge excitonic transition energies are almost unchanged. Compared to other transition-metal chalcogenides, $ReS_2$ exhibits stronger interactions with non-metal adatoms (hydrogen, nitrogen, phosphorus, oxygen, sulfur, fluorine) due to a softer Re–S bonding. The preferential sites for non-metal adatom adsorption are the peak and valley sites of sulfur atoms. Apart from hydrogen, all of other non-metal adatoms can maintain the semiconducting nature of $ReS_2$. Hydrogen adsorption however puts the Fermi level into the conduction band and leads to a semiconductor-metal transition. Nitrogen or phosphorus adsorption creates spin-polarized defect states in the gap and results in half-semiconducting behavior.

It is possible to produce metal-free magnetism via long-range magnetic coupling interactions at low defect concentrations, plus a tunable band-gap. In other materials, the magnetic moment arises from the d-electrons of transition-metal atoms. In $ReS_2$, magnetism based upon the sp states of non-metal elements is cause of metal-free magnetism.

The large-scale fabrication of high-quality two-dimensional nanomaterials remains elusive. An important technique in this regard promises to be space-confined vapor deposition. Using this method, high-quality ultra-thin two-dimensional materials such as

Materials Research Forum LLC
doi: http://dx.doi.org/10.21741/9781945291920

large-area graphene and boron nitride, as well as $ReS_2/ReSe_2$, $HfS_2$, pyramid-structured multilayer $MoS_2$, $Bi_2Se_3$ and $Bi_2Te_3$ have been successfully produced[12]. By using van der Waals epitaxial growth substrates such as mica, the patterned growth of two-dimensional nanomaterials can moreover be readily achieved via surface-induced growth. The choice of a suitable precursor is essential for the chemical vapor deposition of low lattice-symmetry decoupled-interlayer materials because of the latters' tendency to undergo anisotropic and out-of-plane growth, leading to the appearance of thick flakes and dendritic or flower-like structures. The use of volatile $ReO_3$ powder has been found to be advantageous because of its relatively low melting point, but it is also unstable and tends to decompose into $Re_2O_7$ and $ReO_2$ at 400C. The highly volatile $Re_2O_7$ then provides excess vapor and multiple nucleation sites, provoking the growth of many thick 3-dimensional flower-like $ReS_2$ flakes on the substrate. This can be countered by using a confined space and constructing a micro-reactor between two mica plates with $ReO_3$ powder as the precursor. The use of mica reduces the barrier energy for atomic migration in the in-plane direction and promotes a van der Waals epitaxial-growth mechanism. There are however marked differences between the materials deposited on the two mica plates. The outer faces of the mica tend to be covered with thick irregular product while the inner faces tend to be covered with large-scale films consisting of hexagonal monocrystalline grains having domain sizes of up to 60μm. A continuous and uniform monolayer can be obtained by increasing the growth time. The resultant crystals have smooth surfaces with a thickness of about 0.73nm. The small space between the two mica substrates exerts a confining effect which decreases the concentrations of the reactants and hence decreases the nucleation density and growth rate of the product. If the lower mica substrate is partly covered with a second mica plate, the $ReS_2$ which is grown in the 3 regions is quite different in appearance. Growth of the disulfide on mica is surface-dominated and the weak van der Waals interaction between the mica surface and adatoms acts has a surface confinement effect during epitaxial growth. This then causes the growth of films which lie flat on the mica surface, thus suppressing out-of-plane growth.

A rather direct method is to form large rhenium disulfide films via the sulfurization of rhenium films deposited onto c-plane sapphire substrates[13]. A 6mm x 10mm [001]-oriented monocrystalline film with well-defined sharp interface could be produced by sulfurizing rhenium films at 1100C. Incomplete sulfurization and film degradation were observed below and above the 1100C sulfurization temperature. A two-fold symmetry of the monocrystalline in-plane structure, comprising Re–Re bonds together with Re–S bonds, indicated that a distorted 1T structure was the most stable atomic arrangement for rhenium disulfide. When the sulfur/rhenium ratio was equal to, or slightly lower than two, characteristic Raman vibrational modes having the narrowest line-widths were observed. A typical absorption peak was detected at 1.5eV.

One method for the layer-controlled wafer-scale (7cm x 2cm) preparation of rhenium disulfide films of high uniformity and continuity is chemical vapor deposition[14]. Direct synthesis of rhenium disulfide on a transparent flexible glass substrate at low temperatures (450C) can be achieved without the aid of a catalyst or plasma enhancement[15]. Field effect transistors with as-grown rhenium disulfide films on flexible glass typically exhibit n-type behavior: with a low threshold voltage of 0.75V, a high on-off ratio of $10^5$, a low sub-threshold swing of 260mV/decade and a mobility of 0.13cm$^2$/Vs. Direct synthesis of rhenium disulfide films on flexible glass can thus provide a basis for the large-area transfer-free fabrication of high-quality transparent flexible electronic devices.

The direct synthesis of monolayer and multilayer rhenium disulfide by chemical vapor deposition at a low temperature of 450C has been reported. Detailed characterization of the material was performed using various spectroscopic and microscopic methods. Initial field-effect transistor characteristics have been evaluated furthermore, thus highlighting its potential for use as an n-type semiconductor. A systematic study was made of the chemical vapor deposition growth of continuous bilayer rhenium disulfide film and of monocrystalline hexagonal rhenium disulfide flakes[16]. A high-performance photo-detector, based upon a rhenium disulfide flake, exhibited a response of 604/W and a response time of 2ms. A film-based rhenium disulfide photo-detector exhibited a weaker performance than that of a flake-based one, but still had a much faster response time than did those of other reported chemical vapor deposited rhenium disulfide-based photo-detectors.

A proprietary method[17] for preparing rhenium disulfide thin films via chemical vapor deposition involves forming a two-element eutectoid alloy by mixing rhenium and tellurium powders. This alloy then acts as a rhenium source while powdered sulfur is used as the other source. Mica is used as the growth substrate, and two-dimensional rhenium disulfide thin-film occurs in an argon atmosphere at 500 to 900C. As compared with other rhenium disulfide preparation methods, the equipment requirements are modest, the operating process is simple, the reaction temperature is low and the growth efficiency is high. The lattice quality of the product is high, and the technique reliable, producing large-area high-quality product with layer-number control. Another proprietary method[18] involves the *in vacuo* placing of powdered sulfur, powdered rhenium disulfide and a target substrate at the front of a vacuum tube furnace, at the center of the furnace and at the back of the furnace, respectively. Argon is then introduced so as to serve as a working gas. The powdered sulfur is volatilized so as to form a protective atmosphere and the temperature and argon flow are controlled such that the rhenium disulfide is evaporated at the higher temperature and carried to the target substrate via the argon. The rhenium disulfide molecules are thereby deposited onto the target substrate. Following deposition, the product is rapidly cooled, and rhenium disulfide film prepared. Although a batch process, the method is simple in operation, high in efficiency and good with regard to reproducibility. The resultant disulfide is high in mass, and the film thickness and area are controllable. Yet another such method[19] for rhenium diselenide nano-sheet preparation involves reducing a rhenium source and a selenium source using hydroxylammonium chloride in a hydrothermal environment. This produces rhenium diselenide nano-particles for further treatment and avoids the disadvantages of competing preparation processes such as slowness, expense and complexity.

The low lattice symmetry and interlayer decoupling of rhenium disulfide make asymmetrical and out-of-plane growth quite probable, and mainly thick flakes, dendritic forms and flower-like structures have typically been obtained. An approach based upon space-confined epitaxial growth has been developed for the controlled synthesis of rhenium disulfide films[20]. Using this approach, large-area high-quality disulfide films of uniform monolayer thickness are grown onto a mica substrate. The weak van der Waals interaction between the mica surface and disulfide clusters, which favors surface-confined growth and avoids out-of-plane growth, is pivotal in growing rhenium disulfide of uniform monolayer thickness. Morphological changes in the disulfide as a function of growth temperature reveal that asymmetrical growth can be suppressed at relatively low temperatures. A rhenium disulfide field-effect transistor has exhibited a current on/off ratio of $10^6$ and an electron mobility of up to $40 cm^2/Vs$, together with a remarkable photoresponse of 12A/W. This technique provides a means for controlling the unusual growth behavior of low lattice-symmetry two-dimensional layered materials.

A tellurium-assisted chemical vapor deposition method for the large-scale synthesis of high-quality monolayer rhenium disulfide on a mica substrate has been demonstrated[21]. Large-area monolayer rhenium disulfide can be grown in a single-temperature tubular furnace equipped with a 1in-diameter quartz tube under atmospheric pressure. Rhenium and tellurium powders are mixed in the weight ratio of 1:6, and placed in a ceramic boat in the center of the tube furnace. Freshly-cleaved fluorophlogopite mica is then put on the ceramic boat. Typical growth conditions are a carrier gas (argon) flow-rate of 80sccm, a temperature of 700C and a process time of 600s. Atomic force microscopy reveals that the rhenium disulfide has a thickness of 0.7nm; essentially a monolayer. Raman spectra indicate that high-quality rhenium disulfide can be obtained at above 600C.

Materials Research Forum LLC
doi: http://dx.doi.org/10.21741/9781945291920

The venerable chemical vapor transport technique for single-crystal growth has been extended from growing three-dimensional crystals, to preparing two-dimensional atomic layers, by fine-tuning the growth kinetics[22]. Individual monocrystalline flakes and continuous monolayer films have been obtained via chemical vapor transport at 300 to 600C. In order to pursue this direct growth of 2-dimensional semiconductors by reducing the dimensionality of the product, the strategy was to modify the reaction ampoule and slow the growth kinetics. The monolayers can exhibit a high crystallinity and a comparable mobility to those of mechanically exfoliated samples, as revealed by means of atomic-resolution microscopic imaging and electrical transport measurements.

Polycrystalline thin films of rhenium disulfide have been deposited by means of aerosol-assisted chemical vapour deposition at 475C[23]. Such films are promising candidates as models for the incorporation of technetium into transition-metal dichalcogenides as a means of immobilisation during nuclear waste processing. Bottom-up (aerosol-assisted chemical vapor deposition) and top-down (liquid-phase exfoliation) have also been used in tandem in order to produce colloids of few-layer rhenium disulfide in N-methyl pyrrolidone[24]. This processing route is a potentially reliable means for the manufacture of two-dimensional materials.

Isopycnic density gradient ultracentrifugation is routinely used to sort nanomaterials but the usual density-gradient medium, iodixanol, has a maximum density which prevents the use of ultracentrifugation for sorting high-density nanomaterials. This limit can be overcome by adding cesium chloride to iodixanol in order to increase its maximum buoyant density to a point where the high-density rhenium disulfide can be sorted in a layer-by-layer manner[25]. The resultant aqueous rhenium disulfide dispersions exhibit photoluminescence at about 1.5eV; consistent with its direct band-gap semiconductor nature. Photocurrent measurements of thin films formed from solution-processed rhenium disulfide exhibit a spectral response which is consistent with optical absorption and photoluminescence data.

Large-area few-layer rhenium disulfide has been grown onto $SiO_2/Si$ substrates by physical vapour deposition with rhenium disulfide powder as the source material. A ceramic boat containing 30mg of $ReS_2$ powder was placed at the center of a 1in quartz tube and a $SiO_2(280nm)/Si$ substrate was positioned far downstream from the powder. The quartz tube was first pumped down to vacuum, and argon carrier gas (50sccm) was then introduced. The furnace was heated to 900C within 1h and maintained at that temperature for 1h in order to grow the $ReS_2$ film. The furnace was then cooled, and the product annealed in order to improve further the crystalline quality. Subsequently, 400mg of sulfur powder - placed upstream in the furnace - was heated to 165C for 300s and held for 1.5h. The $ReS_2$ at the center of the furnace was annealed at 800C for 1h at a heating-rate of 20C/min under an argon flux of 100sccm at ambient pressure. Optical photographs of the as-grown film showed that the color of the continuous film was pink and that the film was of the order of a centimetre in size. The pink color suggested that the thickness of the film amounted to a few layers. The size could be much greater if the quartz tube used was larger. The film was continuous and homogenous across large areas of the $SiO_2$/Si substrate. The thickness of the film was measured using atomic force microscopy. The disulfide film on the $SiO_2$/Si substrate was homogenous, with a thickness of 2.30nm; equivalent to about 3 monolayers. Raman spectra for the film were collected using 514nm laser excitation. Typical characteristic Raman peaks, at 162 and 213/cm, corresponded to in-plane ($E_{2g}$) and out-of-plane ($A_{1g}$) vibrational modes, respectively. There was also a series of other Raman peaks, between 100 and 400/cm, resulting from the unique asymmetry of the distorted 1T $ReS_2$ structure. X-ray photo-electron spectra between 0 and 600eV comprised 5 elements, including rhenium, sulfur, carbon, silicon and oxygen. The silicon and oxygen peaks originated from the $SiO_2$/Si substrate. The carbon peak was due to contaminants. As-synthesized film exhibited two rhenium 4f peaks at 40.9 and 43.3eV; corresponding to the $4f_{7/2}$ and $4f_{5/2}$ binding energies of $Re^{4+}$, respectively. Two peaks, at 161.4 and 162.6eV respectively, corresponded to the $2p_{3/2}$ and $2p_{1/2}$ binding energies of $S^{2-}$. The atomic ratio of sulfur and rhenium was roughly 2.0, thus demonstrating that the physical vapor-deposited disulfide film was stoichiometric. The surface of the film was clean and continuous at the micron scale. Transmission electron microscopic images suggested that the as-grown film was very thin and was continuous at the nm scale. Dark-field transmission electron microscopic images indicated that the average grain size was 250nm. High-resolution transmission electron microscopy revealed clear lattice fringes from the edge of the film, thus demonstrating that a lattice distance of $d_{(100)}$ of 0.61nm corresponded to the interlayer spacing. The interplanar distance of 0.24nm corresponded to the ($\bar{1}22$) lattice plane of a $ReS_2$ crystal. High-resolution transmission electron microscopy confirmed that the thickness of the disulfide amounted to a few monolayers. Overall, the clear lattice fringes seen in high-resolution transmission electron micrographs confirm that the rhenium disulfide is few-layered and of high crystalline quality[26]. This suggested that physical vapor deposition is a reliable means for synthesizing wafer-scale rhenium disulfide films.

It is interesting to compare the details of the above method with those of a very similar one. Highly monocrystalline two-dimensional rhenium disulfide nanostructures can be obtained by direct chemical reaction between rhenium atoms, originating from the decomposition of $ReCl_3$, and sulfur atoms on a $SiO_2$ substrate under flowing helium at 450C[27]. The technique used to obtain nanosheets with a vertical orientation at ambient pressures is to react rhenium and sulfur and let the product crystallize on a $SiO_2$/Si substrate. This is simple method for the preparation of few-layer nanostructures at low temperatures and in high densities. A quartz tube is first flushed with 6N-purity helium carrier gas at a flow-rate of 500sccm for 300s. A $SiO_2$(200nm)/Si(001) substrate is then placed face-down over a quartz boat containing rhenium trichloride at the centre of a tube furnace. Another quartz boat containing sulfur powder is then placed 15cm up-stream from the substrate, which is near to the outer edge of the hot zone of the furnace. The latter is heated 450C at the rate of 28C/min under a helium flow of 10sccm and held for 20min at 450C. During the growth process, the temperature around the sulfur powder is kept at about 180C. When growth is complete, the furnace is turned off and allowed to cool to room temperature. Scanning electron microscopy and atomic force microscopy then typically show that a high density of nanosheets is present, and that they are mainly oriented vertically on the substrate. Their sizes range up to 1μm, with a thickness of less than 10nm. The edges of the nanosheets are generally connected to other out-of-plane nanosheets, but asymmetrical crystal growth and out-of-plane growth may favour the creation of the distorted 1T structure and a weak interlayer strength. The vertical orientation of the nanosheets is itself attributed in part to a weak interaction between the disulfide and the $SiO_2$ substrate. That in turn is attributed to the positioning of the substrate with respect to the rhenium source. The height of a nanosheet is 6.6nm, thus indicating the presence of some 4 or 5 layers of nanosheets. It is found that, although nanosheet growth can occur at 350C, the nanosheets are then too small and posses an ill-defined morphology. As the chosen growth temperature is increased from 400 to 550C, the size of the nanosheets increases due to more rapid growth at higher temperatures. Growth at 550C produces a much higher density than growth at 450C, and the product is in the form of a continuous thin film which covers the entire substrate. Growth at 600C produces very thick flake-like nanostructures. X-ray diffraction studies of nanosheets grown at 450C show that there are only 4 prominent peaks: 14.80, 29.60, 44.88 and 61.82°. The most intense peak, at 14.80°, is assigned to the (001) plane of triclinic $ReS_2$, and the lattice constant of the c-axis is estimated to be 0.6282nm. The other peaks are those for the (002), (003) and (004) planes of triclinic $ReS_2$. The fact that the (00l) reflection is present while all of the other (hkl) reflections are absent implies that the crystal structure is well-oriented along the c-axis. The sharp characteristic peaks also confirm the degree of crystallinity of the two-dimensional layered structures. The Raman spectrum of as-grown nanosheets under 632.8nm laser excitation is consistent with previously reported values for few-layer disulfide nanostructures, and the most intense peak (150/cm) is a typical first-order in-plane $E_{2g}$ vibrational mode while a peak at 210/cm is a characteristic Raman mode for out-of-plane ($A_{1g}$) vibration. Other Raman modes, between 100 and 400/cm, are attributed to symmetry-splitting in the distorted 1T structure. Satellite Raman peaks between 300 and 400/cm originated from second-order Raman modes. Analysis revealed that silicon and oxygen contents originated from the $SiO_2$ substrate while carbon originated from the environment. Predominant peaks at binding energies of 42.6 and 45.1eV are assigned to $Re^{4+}$ 4f states. The characteristic sulfur $2p_{3/2}$ binding energy of the divalent ion oxidation state is located at 163.1eV. The region of the core-level sulfur 2p peaks is relatively broad and a peak at 164.2eV, assigned to the sulfur $2p_{1/2}$ state, is also a characteristic binding energy of the divalent ion. The rhenium/sulfur atomic ratio is estimated to be 1:1.98; showing that the as-grown nanosheets are highly stoichiometric. Z-contrast high-angle annular dark field scanning transmission electron microscopy indicates the formation of differing numbers of layered nanosheets. It is concluded that, at the 450C growth temperature, the $ReCl_3$ source can decompose into rhenium and $Cl_2$. The rhenium atoms can then directly react with sulfur atoms on the $SiO_2$ substrate to form tiny $ReS_2$ nuclei. The continuous feeding of rhenium and sulfur atoms to nucleation sites near to these nuclei then lead to a two-dimensional layered disulfide structure. The supply of rhenium atoms from the gaseous phase of $ReCl_3$ is pivotal in obtaining nanosheets possessing high crystallinity and high density. If $ReO_3$ is used as the rhenium source for chemical vapor deposition, the resultant disulfide nanosheets obtained at about 450C are highly amorphous and in much lower density, due to the formation of other rhenium oxides such as $Re_2O_7$ and $ReO_2$. In the present method, as the rhenium atoms are obtained from the direct decomposition of $ReCl_3$ in flowing helium, the formation of oxides is not likely. This approach to producing rhenium disulfide nanosheets is simple, cost-effective, and versatile.

It has been demonstrated[28] that the twinned growth relationship between two two-dimensional materials can be used to construct vertically-stacked heterostructures. One example involved the 100% overlap of two transition metal dichalcogenide layers, $ReS_2/WS_2$, where the crystal size of the stacked structure was an order-of-magnitude larger than that previously reported for such materials. In order to achieve one-step growth of $ReS_2/WS_2$ twinned vertically-stacked heterostructures, a re-solidified gold substrate was used in which rhenium and tungsten had been dissolved. A piece of gold wire was then placed on a tungsten-rhenium alloy foil. Under an $Ar/H_2$ atmosphere, the gold was then spread evenly over the whole foil by annealing at 1100C for about 600s. This also permitted the rhenium and tungsten atoms to diffuse into the gold lattice. The temperature was then decreased to 900C for chemical vapor deposition growth. The simultaneous growth of $ReS_2/WS_2$ layers to form a vertically stacked heterostructure upon introducing $H_2S$ into the system for 600s. These twinned heterostructures exhibited a better hydrogen evolution reaction activity as compared with pure $WS_2$, and therefore offered great potential as catalytic materials. The simplicity of twinned growth could be exploited in expanding the fabrication of other heterostructures or two-dimensional superlattices in the field of two-dimensional van der Waals heterostructures.

A highly controllable atomic layer deposition technique[29] has been used to deposit $ReS_2$ thin films onto large (5cm x 5cm) substrates at 120 to 500C. Extensive characterization using field emission scanning electron microscopy, energy-dispersive X-ray spectroscopy, grazing-incidence X-ray diffractometry, atomic force microscopy, focused ion beam microscopy, transmission electron microscopy, X-ray photo-electron spectroscopy and time-of-flight elastic recoil detection analysis confirmed that this $ReS_2$ atomic layer deposition process increased the potential of the material for applications beyond planar-structure architectures and could be scaled-up to industrial levels. Mechanical exfoliation is a good method for studying the suitability of two-dimensional materials for applications but, for large-scale production, there is a need to deposit such materials onto large-area substrates using vapor phase techniques. Modified chemical vapor deposition is the best method for the deposition of thin films. Atomic layer deposition is known for its ability to ensure area-uniformity plus simple and accurate film-thickness and composition control, superior reproducibility and scalability. Self-limiting surface reactions, produced by alternating precursor pulses, guarantee the above advantages much more so than does chemical vapor deposition. Rhenium disulfide films were grown by using conventional metal halide and $H_2S$-based atomic-layer deposition using deposition temperatures of up to 500C and 5cm x 5cm substrates. Conformal coating of a complex three-dimensional structure was also achieved. By controlling the film thickness and the deposition temperature, either smooth well-oriented films or edge-exposing flakes standing out from the substrate plane could be obtained. The deposition-temperature dependence of atomic-layer deposited film growth was investigated using *in situ* $Al_2O_3$-coated Si(100) substrates. Successful film growth was possible between 120 and 500C, and was limited only by the temperature (110C) required to sublimate $ReCl_3$ and the maximum temperature (500C) of the atomic layer deposition reactor glass tubes The growth rate increased from about 0.3Å/cycle to nearly 0.9Å/cycle between 120 and 200C, and the film structure changed from X-ray amorphous to crystalline. The growth rate was 0.8 to 0.9Å/cycle at 200 to 300C, and then steadily decreased to about 0.2Å/cycle at 450C. This decrease in the growth rate was related to a decrease in the intensity of a wide peak at 32° to 34° and to the appearance of a (001) orientation in X-ray diffractometry grazing incidence patterns. Field emission scanning electron microscopy revealed that, at up to 300C, the crystals were oriented parallel to the substrate. At 350C and above, the plate-like crystals were increasingly randomly oriented with respect to the substrate. The calculated nominal film thickness and growth rate were calculated under the assumption that the films were smooth, uniform and continuous and had a density of $7g/cm^3$; the bulk density of $ReS_2$ being $7.6g/cm^3$. The 80 to 90nm thick films which were grown at 200 to 250C were very smooth and had a surface roughness of only 2nm. This increased to nearly 6nm for film which was grown at 300C. The underlying $Al_2O_3$ film could not be confidently distinguished from the $ReS_2$ and silicon substrate, using time-of-flight elastic recoil detection analysis, because of the upward orientation of the $ReS_2$ crystals and because of the limited mass resolution separation of aluminium and silicon. A roughly 7nm thickness for the $Al_2O_3$ film was deducted by assuming the presence of stoichiometric amounts of aluminum and oxygen in the latter film. According to the calculated sulfur/rhenium ratios of 1.8 to 2.1, crystalline films which were grown at 200C or above appeared to be stoichiometric $ReS_2$ (2.0) rather than $Re_2S_7$ (3.5). Amorphous films which were deposited at 150C or below contained large amounts (18 to 28at%) of oxygen. This suggested that the films had oxidized following deposition, or that the growth was affected by volatile oxychlorides and oxides which formed during the loading of precursors. The oxygen content was 3 to 4at% for smooth crystalline films, and decreased to 1at% or less for vertical $ReS_2$ crystals which had been grown at higher temperatures. According to X-ray photo-electron spectroscopic data, most of the oxygen on the film surfaces was in the form of hydroxyl groups. This suggested that the oxygen contamination of 4 to 9at% on the crystalline film surfaces arose from exposure to the ambient atmosphere following deposition. The film growth rates were investigated at 400C by varying the $ReCl_5$ and $H_2S$ pulse lengths. At this temperature, the films consisted of randomly oriented $ReS_2$ crystals. The saturation of the growth rate which is typical of atomic-layer deposition was not observed. Increasing the $ReCl_5$ pulse length decreased the nominal growth rate, while increasing the $H_2S$ pulse length increased the nominal growth rate. The absence of saturation of the growth rate was attributed to the unique film morphology. The precursors had opposing effects on the film morphology: increasing the $ReCl_5$ pulse length tended to lead to fewer vertical $ReS_2$ flakes, while a $H_2S$ pulse increment resulted in areal densification of the flakes on the surface. Although no typical growth-rate saturation was observed, the nominal film thickness increased linearly with increasing number of deposition cycles, as expected of an atomic layer deposition process. The film growth progressed from X-ray amorphous to crystalline with a visible (001) orientation after 200 deposition cycles. Additional orientations appeared in the films which were grown using more than 1500 cycles. Anorthic $ReS_2$ has a complicated X-ray diffraction pattern, with several neighboring orientations. Specific peaks from the films could therefore not be accurately assigned. Randomly oriented crystalline $ReS_2$ film with a 32nm nominal thickness and a surface roughness of 23nm consisted of a roughly 10nm-thick uniform film which was oriented

Materials Research Forum LLC

doi: http://dx.doi.org/10.21741/9781945291920

along the substrate, plus flakes of similar thickness which emerged - from the uniform film - in directions oriented away from the substrate plane. The interlayer spacing which was deduced from transmission electron microscopic images was 6.3Å. The interlayer spacings which were deduced for films grown at 300, 400 and 500C were 6.5, 6.3 and 6.5Å, respectively. These values were in good agreement with the reported interlayer spacings of 6.0 to 6.9Å. The film growth of $ReS_2$ also occurred on a three-dimensional trench structure, further demonstrating the utility of atomic layer deposition.

Another proposed method[30] involves using ammonium perrhenate as a rhenium source and thiourea as a sulfur source, with hydroxylamine hydrochloride as a reducing agent. These three reactants are dissolved in water and the chemical reaction is controlled by regulating the temperature and time. The product is then cleaned and dried before producing rhenium disulfide nanosheets. This method promises to be simple in operation and high in efficiency, with good reproducibility and low cost. High-quality rhenium disulfide nanosheets are expected to be available in large batches.

Few-layer rhenium disulfide nanosheets have been directly nucleated and grown onto the tube walls of carbon nanotubes by using a simple one-pot hydrothermal method[31]. As compared to the plain disulfide, a $ReS_2$/nanotube composite anode offers a markedly enhanced electrochemical performance, with the very high capacity of 1048mAh/g at 0.2C and a high-capacity retention of 93.6% after 100 cycles at 0.5C. These figures are much better than those for the plain disulfide. This marked enhancement is attributed largely to the unique architecture. That is, the carbon nanotubes guarantee the presence of a highly conductive porous inter-network, and also ensure intimate contact with the disulfide. This then permits easy electrolyte infiltration, efficient electron transfer and ionic diffusion. This synthesis method can be extended to the synthesis of other two-dimensional semiconductor-based composites for energy storage and catalysis.

High-quality single crystals of rhenium disulfide have been synthesized by using a modified Bridgman method which avoids the use of a halogen transport agent[32]. X-ray diffraction and electron microscopy confirm the presence of a distorted triclinic 1T′ structure, and reveal a lack of Bernal stacking. Photoluminescence measurements indicate a layer-independent band-gap of 1.51eV, with an increased photoluminescence intensity arising from thicker flakes. This confirms that interlayer coupling is therefore negligible in this material. The lower degree of background doping which this crystal growth process entails leads to a high field-effect mobility of $79cm^2/Vs$, as deduced from field effect transistor structures produced from exfoliated flakes. This work again demonstrates that such chalcogenides are promising two-dimensional materials for optoelectronic devices, and do not have to be monolayer thin flakes in order to offer a direct band-gap.

Rhenium disulfide onion-like nanoparticles can be prepared by high-temperature metal-organic chemical vapor deposition, beginning with $Re_2(CO)_{10}$ and elemental sulfur[33]. The reaction is carried out in two stages, in which the intermediate product of amorphous rhenium disulfide nanoparticles - formed by the high-temperature reaction of rhenium and sulfur in the first part of the reaction – is isolated. This is then converted into onion-like rhenium disulfide nanoparticles during a separate annealing step. Analysis of the reaction product by using X-ray diffraction and high-resolution transmission electron microscopy, combined with energy-dispersive X-ray spectroscopy, guides the optimization of the reaction so that onion-like structures alone are formed.

High yields of dense microspheres consisting of poorly crystallized rhenium disulfide embedded in carbon have been synthesized by using a simple one-pot solvothermal route: reacting dirhenium decacarbonyl, elemental sulfur and an aromatic solvent such as benzene, toluene or p-xylene for 24h at 180C[34]. X-ray diffraction, scanning electron microscopy, high-resolution transmission electron microscopy, energy-dispersive X-ray spectroscopy, Raman spectroscopy and Fourier transform infra-red spectroscopy all confirmed that the resultant microspheres were dense, with smooth surfaces, and had average diameters of between 0.79 and 1.40µm. They consisted of rhenium disulfide sheet-like structures with 4.5 to 9.8wt% of structural amorphous carbon, which was retained as a texture-stabilizer following calcination at 800C. When the synthesis is performed using isopropanol and cyclohexane, the products are agglomerated grains and botryoidal quasi-spherical particles, respectively.

Another approach to the synthesis of highly de-stacked rhenium disulfide layers embedded in amorphous carbon is the thermal decomposition of a tetra-octylammonium perrhenate precursor[35]. Ammonium perrhenate (6.0mmol) was dissolved in 30ml of water and the solution was added to 6.0mmol of tetra-octylammonium bromide which had previously been dissolved in a solution of 10ml water plus 5ml ethanol at 70C. The precipitate, formed by ionic exchange, was washed with distilled water and allowed to dry at room temperature. In order to obtain rhenium disulfide monolayers, the precursor was decomposed in a reducing atmosphere. The octylammonium $ReO_4$ was heated to 400C at 2C/min under flowing $15\%H_2S$ in hydrogen and maintained thus for 4h before being allowed to cool to room temperature. X-ray diffraction, scanning electron microscopic, scanning transmission electron microscopic, energy-dispersive X-ray spectroscopic and X-ray photo-electron spectroscopic investigations of the product all confirmed the formation of rhenium disulfide. The product was composed of microscopic irregular grains having diameters of between 0.1 and 2.0µm. Closer examination showed that the grains appeared to be agglomerations of sub-micrometric platelets. Scanning transmission electron microscopy revealed that there was a higher proportion of poorly-stacked nanosheets having a random orientation. There were a few thin arrangements of stacked layers along the c-axis with an average interplanar distance of 0.62nm, as indicated by diffuse diffraction rings in the corresponding selected area electron diffraction patterns. The layer length was deduced from a statistical analysis that was based upon 400 layers in various parts of the images. The longitudinal slab length of the layers was about 4.81nm. The composition of the product included rhenium, sulfur and carbon plus an appreciable amount of oxygen. The sulfur/rhenium ratio was about 1:1.55 and the carbon and oxygen contents were about 35.2 and 31.4at%, respectively. X-ray photo-electron spectroscopic peaks at 41.9 and 162.7eV were assigned to rhenium $4f_{4/7}$ and sulfur $2p_{3/2}$ electrons, respectively. The rhenium spectrum also revealed the presence of other photo-generated rhenium f electrons having higher relative binding energies. Deconvolution of 3 additional doublets on the rhenium curve revealed the presence of oxides on the sample surface: $ReO_2$ (43.5eV), $ReO_3$ (45.7eV), $Re_2O_7$ (46.6eV). In the sulfur spectrum, emissions (169.2eV) which could be attributed to sulfates were observed. The carbon spectrum was typical of carbonaceous materials, in which the emissions can be assigned to aromatic C-C (284eV), aliphatic C-C(285eV) and hydroxyl carbon C-O (286.1eV). Overall, the compound is found to be in the form of single layers, with a small proportion of few-layer arrangements, embedded in amorphous carbon. X-ray diffraction, ultraviolet-visible diffuse reflectance, and thermogravimetry analysis show that the perrhenate ions are widely separated from one another within the matrix of the organic cations; forming an inorganic-organic salt. This special arrangement of the disulfide layers is of possible use as a heterogeneous catalyst, due to the high proportion of edge-sites.

Substitutional doping of transition-metal dichalcogenide two-dimensional materials such as rhenium disulfide is effective in tuning intrinsic properties such as the band-gap, transport characteristics and magnetism. Substitutional doping of monolayer rhenium disulfide with molybdenum has been achieved by chemical vapor deposition[36]. Scanning transmission electron microscopy confirms that molybdenum atoms are successfully doped into the disulfide by substitutionally replacing rhenium atoms in the lattice. Electrical measurements reveal a degenerate p-type semiconducting behavior in molybdenum-doped rhenium disulfide field effect transistors; in agreement with density functional theory calculations. A p-n diode device, based upon doped rhenium disulfide and a disulfide homojunction exhibited a gate-tunable current rectification behavior. The maximum rectification ratio could reach 150 at -2/+2V.

There remains the need for a simple general method for the preparation of transition-metal dichalcogenide nanodots. One candidate method[37] has been used to prepare $ReS_2$ (or the molybdenum, tungsten, tantalum sulfide and selenide) nanodots from bulk crystals by combining grinding with sonification. The nanodots could be easily separated from N-methyl-2-pyrrolidone by subsequent treatment with n-hexane and chloroform. All of those transition-metal dichalcogenide nanodots which were smaller than 10nm had a narrow size distribution, with a high dispersion in solution. As a test-application, memory devices were created by using transition-metal dichalcogenide nanodots, mixed with polyvinylpyrrolidone, as active layers. These possessed non-volatile write-once-read-many capabilities.

**Molecular Structure**

Most transition-metal chalcogenides possess a graphene-like hexagonal crystal structure in which the atoms are sandwiched between layers of chalcogen atoms. The chalcogen layers are stacked above one another as an hexagonal phase made up of prismatic holes for metal atoms, or as a trigonal phase made up of octahedral holes for metals. Structural polytypism is thus possible among group-VI transition-metal chalcogenides. The common phases of group-VI transition-metal chalcogenides include 2H, 1T and 3R phases for bulk crystals, and 1H and 1T for monolayers; the T-phases tending to be metastable. The important feature of $ReS_2$, as compared with $MX_2$ and $MX_3$ chalcogenides involving metals from groups IV to VI, is its triclinic structure and the in-plane structural, optical, electrical and mechanical anisotropies existing along the [010] b-axis.

Due to metal-metal bonding, the unit cell of $ReS_2$ is doubled, leading to a composition containing 4 rhenium and 8 sulfur atoms. In the crystal, $Re_4$ units form parallel one-dimensional chains, in monolayers, which can distort or break the hexagonal symmetry. Unlike transition-metal chalcogenides such as $MoS_2$ which have a high-symmetry 1H-phase trigonal prismatic crystal and $D_{3h}$ point-group symmetry, or 2-layer hexagonal and three-layer rhombohedral structure, $ReS_2$ exhibits Peierls distortion of the conventional 1T-phase of an octahedral crystal. Due to the Peierls distortion, the formation of rhenium chains breaks the hexagonal symmetry. The rhenium and chalcogen atoms in a unit cell are then forced into the same plane and produce an in-plane and out-of-plane anisotropy along the lattice vectors. The asymmetrical unit of $ReS_2$ comprises two $Re^{4+}$ ions and four $S^{2-}$ ions. The sulfur-atom layers are hexagonally close-packed and are stacked along the a-axis; nearly parallel to the bc-plane of the unit cell. The rhenium atoms occupy the octahedral sites between every other pair of hexagonal close-packed layer of sulfur atoms. Each rhenium atom is coordinated with 6 sulfur atoms in an octahedral geometry, and each trigonal pyramidal sulfur atom is bonded to 3 rhenium atoms. The latter form metal-metal bonds with 3 neighbors in the cation layers, resulting in the formation of a $Re_4$ parallelogram which markedly affects the relative displacements of sulfur atoms within their hexagonal close-packed array. The material is made up of atomic layers, S–Re–S, wherein the rhenium and sulfur atoms are connected by covalent bonds. Adjacent layers are coupled by weak van der Waals forces so as to form bulk crystals. Transition-metal dichalcogenides exist in several structural forms, and are characterised by the differing coordination spheres of the transition-metal atom. The most common forms are trigonal prismatic (2H) and octahedral (1T) coordinations, and are defined by the various stacking orders of the 3 chalcogen–metal–chalcogen atomic planes which make up the individual layers. The 2H-form involves an ABA stacking in which the chalcogen atoms of various atomic planes occupy the same A-type position and are superposed in the direction perpendicular to the layer. This is not the form exhibited by rhenium disulfide. The 1T form instead involves an ABC stacking. The structure of these materials is also defined by the configurations of individual layers in multilayer and bulk samples. Structural distortions can reduce the periodicity, and large distortions can lead to the formation of metal–metal bonds as in the case of 1T-$ReS_2$. Group-VIIB transition-metal dichalcogenides are stable only in the form of the distorted 1T phase, 1T″, which is common to many stable and metastable phases in other transition-metal dichalcogenides. First-principles calculations of the structural origin of 1T″-phase group-VIIB transition-metal dichalcogenides show[38] that a quasi one-dimensional Peierls-like instability is the reason for the transition to a 1T″-phase $ReS_2$ monolayer from the 1T′ phase, another distorted 1T phase.

Looked at from a slightly different viewpoint, each layer of $ReS_2$ resembles the 1T crystal structure with an embedded one-dimensional parallel chain of $Re_4$ clusters within each monolayer. Each rhenium atom has seven valence electrons, plus one dangling electron in the 1T structure. The extra electrons lead to strong covalent bonding between the rhenium atoms. The formation of Re–Re covalent bonds then affects the energy balance of the system. Due to the covalent bonding, the lattice then adopts the distorted 1T where the rhenium atoms of the layer form a zig-zag Re–Re chain. The presence of the chains in each layer makes the energy difference of the layers during sliding of minimal importance. The extra electron on the rhenium atom leads to much lower intralayer polarization, and weakens the interlayer van der Waals interaction. The resultant lack of ordered stacking, plus weak intralayer polarization, are the principal reasons for the almost non-existent interlayer coupling of the layers.

The present material contains rhenium which, having a higher atomic weight than other transition-metals, exhibits a high spin-orbit interaction. Its unit cell is also relatively large and asymmetrical. Monolayers of $ReS_2$ are highly anisotropic and contain both metal-metal and metal-chalcogen bonds. Lattice vibrations play a pivotal role in the special properties of the layered two-dimensional material. The interlayer interaction of its van

Materials Research Foundations **40** (2018)               doi: http://dx.doi.org/10.21741/9781945291920

der Waals structure assists lattice vibration and the appearance of shear and breathing phonon-modes. The two types of active Raman modes in two-dimensional $ReS_2$ are intralayer vibration modes and interlayer vibration modes. The former usually appear at higher frequencies. The interlayer vibration mode can be sub-divided into shear and breathing modes and usually appear at frequencies below 50/cm. They are difficult to observe. The number of Brillouin zone center, $\Gamma$, phonons for a given two-dimensional material is equal to 3 times the number of atoms in the unit cell. The $\Gamma$-point phonons combine acoustic and optical modes, and not all are Raman-active. Thus a $MX_2$ compound with $D_{6h}$ point-group symmetry, and a unit cell containing 6 atoms will have 18 phonon modes that can be described as: $\Gamma = A_{1g} + 2A_{2u} + E_{1g} + 2E_{1u} + 2E_{2g} + E_{2u} + 2B_{2g} + B_{1u}$; $A_{1g}$, $E_{1g}$ and $2E_{2g}$ being Raman-active. The unit cell of $ReS_2$, consisting as it does of 12 atoms, will have 36 vibrational modes and 36 $\Gamma$-point phonon modes that can be written: $\Gamma = 18(A_g + A_u)$. Among these 36 $\Gamma$-point phonon modes, the 18 $A_u$ modes consist of 3 acoustic and 15 infra-red optical modes which are asymmetrical and inactive in Raman spectroscopy. This suggests that there are only 18 Raman modes in $ReS_2$, where the $A_g$ represent out-of-plane vibrational modes, the $E_g$ are in-plane vibrational modes and the $C_p$ are in-plane and out-of-plane coupled modes. All of the Raman active and inactive modes are non-degenerate. There are 4 out-of plane vibrational modes, 6 in-plane vibrational modes and 8 coupled modes. The out-of plane modes at 136.8 and 144.5/cm are assigned to the out-of-plane vibrations of rhenium atoms while the modes at 422.3 and 443.4/cm are due to the out-of-plane vibrations of sulfur atoms. Among the in-plane modes, those at 153.6, 163.4, 218.2 and 238.1/cm are caused by the in-plane vibrations of rhenium atoms while the modes at 308.5 and 312.1/cm are due to the in-plane vibrations of sulfur atoms. The $C_p$ modes are a mixture of in-plane and out-of-plane vibrations of rhenium and sulfur atoms. The difference in the Raman shifts of bulk and monolayer samples is minute, thus suggesting ultra-weak interlayer interactions such that the monolayer could behave like a vibrationally-decoupled bulk sample. Rhenium disulfide has a low-frequency Raman spectra which features 3 clear Raman peaks at 13, 16.5 and 28/cm for 2-layer samples. Those peaks are missing from monolayer samples, thus implying that the peaks originate from vibrations between 2 layers. The Raman peak at 28/cm is assigned to a breathing mode, while the peaks at 13 and 16.5/cm are assigned to parallel and vertical shear modes, respectively. It is notable that the shear modes in $ReS_2$ are non-degenerate, but are degenerate for other two-dimensional materials. The breathing mode is also stronger than the shear modes in $ReS_2$, while the shear modes are stronger than the breathing modes in other two-dimensional materials. The appearance of interlayer phonon modes reflects a marked lattice coupling between the $ReS_2$ layers. The observation of shear modes indicates the existence of a well-defined layer-stacking order in $ReS_2$ crystals because the generation of shear-mode vibrations requires a good atomic registration between neighboring layers.

Density functional theory calculations have been performed using a code package with an ultra-soft pseudopotential. A generalized gradient approximation of Perdew-Burke-Ernzerhof form was used to handle electron exchange correlations and energy cut-offs of 60 and 480Ry were chosen for expansion of the plane-wave basis and charge density, respectively. The isolated two-dimensional layers were represented in the super-cell, with a vacuum layer of 16Å between adjacent layers in order to minimize artificial interlayer interactions. Atomic relaxation was performed until the Hellmann–Feynman amounted to less than 5meV/Å, and Brillouin-zone sampling was carried out over 24 x 13 x 1 and 12 x 12 x 1 k-point grids, respectively, for the 1T′ and 1T″ structures. Two-fold denser k-point grids were used meanwhile to calculate the band structure and the Fermi level. The phonon dispersion was deduced by using density functional perturbation theory with a q-vector grid of 8 x 5 x 1 for the 1T′ structure and 5 x 5 x 1 for the 1T″ structure. In order to identify the nesting vector for a possible charge density wave phase, the Lindhard function was calculated. The virtual crystal approximation method was also used to study variation in the electronic structure of $W_{1-x}Re_xS_2$ (x = 0 to 1). Two half-filled bands in the 1T′-$ReS_2$ produce sharp peaks, in the Lindhard function, which introduce a charge density wave phase with a large band-gap opening. Calculations also show that overlap of the two bands over a broad range of energies leads to a stable charge density wave phase, or a stable 1T″ phase, in group-VIIB transition metal dichalcogenides in the face of compositional variations. This behaviour is in marked contrast to the typical Peierls instability which is driven by a single band. The 1T′ phase has a 1 x 2 super-cell structure with P21/m space-group symmetry; like 1T′ group-VIB transition-metal dichalcogenides. But this phase is dynamically unstable, with an imaginary frequency across the Brillouin zone, which implies the possibility of structural stabilization via cell-doubling along the metal-atomic chain direction. When the atomic structure is relaxed following cell-doubling, without any symmetry constraint, the material changes into the 1T″ phase; a 2 x 2 super-cell which possesses only inversion symmetry. The associated energy gain is 0.29eV per formula unit of the 1T′ phase.

The lattice parameters are similar for the two phases, apart from the Re-Re bond length. Unlike the single length in the 1T′ phase, it splits into three lengths in the 1T″ phase. The 1T′-1T″ transition is therefore characterized by Re-atom rearrangement so as to form diamond-shaped Re-Re bonds. This then leads to a metal-to-insulator transition. Using a simple electron-counting rule, 1T′ with a 1 x 2 super-cell could be expected to be an insulator but is in fact metallic due to two degenerate parabolic bands. The two roughly half-filled bands exhibit a large overlap in energy, producing Fermi wave vectors in both bands. This is similar to the one-dimensional Peierls distortion which occurs when the Fermi wave vector is close to being equal to one quarter of the reciprocal lattice vector. The band-structure of the 1T″ phase exhibits a direct band-gap of 1.43eV. The valence-band maximum is located between the Γ and $K_2$ points and is almost degenerate, with the top within about 0.01eV of the Γ point.

*Table 3 Fractional coordinates of the atoms of the ReS₂ unit cell*

| Atom | x | y | z |
|---|---|---|---|
| $Re_1$ | 0.4925 | 0.0564 | 0.2477 |
| $Re_2$ | 0.5026 | 0.5112 | 0.2974 |
| $S_1$ | 0.2174 | 0.2498 | 0.3676 |
| $S_2$ | 0.2769 | 0.7705 | 0.3819 |
| $S_3$ | 0.7562 | 0.2729 | 0.1178 |
| $S_4$ | 0.6975 | 0.7526 | 0.1169 |

The valence-band maximum occurs at the Γ-point when spin-orbit coupling is included. Band-gap opening in the 1T″ phase is due to band-folding and splitting along the metal-chain direction in the 1T′ phase. The 1T′–1T″ transition mainly affects the band dispersion along certain chain directions in the 1T′ phase, while the band dispersion and its nature perpendicular to the chain direction are rarely affected. The band-gap of the 1T″ is also similar in magnitude to the band width of 1T′. Both quantities are governed essentially by energy-splitting of bonding and anti-bonding states in the inner chain. The cell-doubling in one direction, and the corresponding features of band-dispersion, indicate that the 1T′–1T″ transition is due to Peierls instability. The calculated Lindhard function for the two bands which cross the Fermi level showed that the contribution arising from other bands is negligible. The effect of temperature is felt via the Fermi–Dirac distribution. For a temperature increase from 10 to 300K, the peaks remain sharp while the Lindhard function becomes smooth; implying the existence of a stable charge density wave phase even at high temperatures. An out-of-phase charge density wave is common in quasi one-dimensional systems due to Madelung-type Coulomb coupling between adjacent metallic chains. Decomposition of the Lindhard function into band-by-band interactions revealed the predominant contribution to the sharp peak. Total energy curves, calculated for locations near to the critical point, exhibit the features of a first-order Landau transition which is due to local chemical bonding. It is deduced that the

Rhenium Disulfide                                              Materials Research Forum LLC
Materials Research Foundations **40** (2018)          doi: http://dx.doi.org/10.21741/9781945291920

structural stability of the 1T″ phase of group-VIIB transition-metal dichalcogenides is ensured by two half-filled bands and by local chemical bonding.

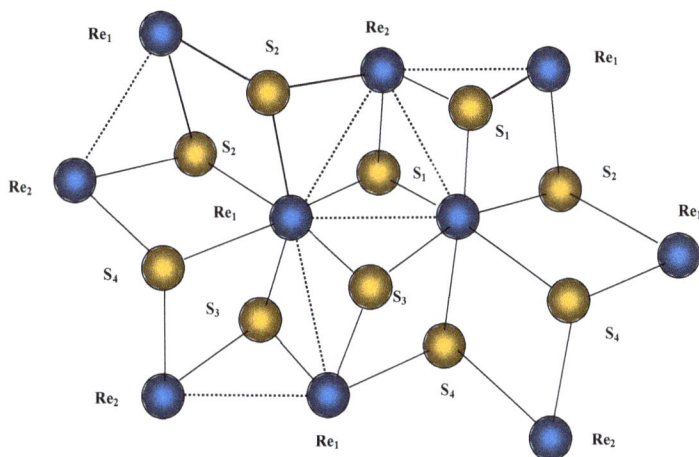

*Figure 2 Asymmetrical cell of ReS₂ viewed in the a-axis direction*

The asymmetrical unit cell of rhenium disulfide (a = 6.417Å, b = 6.510Å, c = 6.461Å, α = 121.10°, β = 88.38°, γ = 106.47°, Z = 4, ρ = 7.581g/cm³) consists of two $Re^{4+}$ and four $S^{2-}$ ions: $Re_1$, $Re_2$, $S_1$, $S_2$, $S_3$, $S_4$ (figure 2, tables 3 and 4). Layers of essentially hexagonal close-packed arrays of sulfur atoms are arranged along the a-axis and are almost parallel to the bc-plane of the unit cell[39]. Rhenium atoms occupy the octahedral sites between alternating pairs of the hexagonal close-packed layers of sulfur atoms. Each rhenium atom is linked to six sulfur atoms in an almost octahedral shape, and each trigonal pyramidal sulfur atom is linked to three rhenium atoms. The involvement of each $d^3$ rhenium atom in the formation of metal-metal bonds to three neighbours in the cation layer produces $Re_4$ parallelograms in which the acute vertices are chain-linked linearly by the 2.895Å rhenium-rhenium bonds. Each $Re_4$ parallelogram is made up of two metal-metal bonded rhenium atoms which share an edge. The Re-Re bonds which form the parallelogram exert a marked effect upon the displacements of the sulfur atoms. The parallelograms are within 1.3° of being coplanar, with the bc-plane of the unit cell, and represent a fictive basal plane with respect to which sulfur-atom displacements can be judged. Such displacements from this basal plane, and into the space between $[ReS_2]_x$

Materials Research Forum LLC
doi: http://dx.doi.org/10.21741/9781945291920

layers is closely related to the number of metal-metal bonds involved. So $S_1$, spanning three metal-metal rhenium atoms is displaced by 1.71Å. Meanwhile $S_3$, spanning two rhenium atoms, is displaced by 1.58Å. Similarly $S_2$, spanning a single metal-metal bond, is displaced by 1.41Å and $S_4$, spanning zero bonds is displaced by 1.14Å. The effect of these differing displacements is to impart a rippled shape to the layer of sulfur atoms.

The high-pressure behavior of rhenium disulfide has been investigated (figures 3 and 4) at up to 51.0GPa by means of *in situ* synchrotron X-ray diffraction in a diamond anvil cell at room temperature[40]. It was found that the rhenium disulfide triclinic phase is stable at up to 11.3GPa. At this point, the disulfide transforms into a new high-pressure phase, which has been tentatively identified as being an hexagonal lattice with a P6m2 space group. The high-pressure phase was stable up to the highest pressure studied, and could not be preserved during decompression. The compressibility of the triclinic phase is anisotropic; being more compressive along interlayer directions than along intralayer directions. This demonstrates the effect of the weak interlayer van der Waals interactions and of the strong intralayer covalent bonds.

*Table 4 Bond lengths in rhenium disulfide*

| Bond | Length (Å) |
|---|---|
| $Re_1$-$Re_2$ | 2.790 |
| $Re_1$-$S_1$ | 2.341 |
| $Re_1$-$S_3$ | 2.311 |
| $Re_2$-$S_1$ | 2.331 |
| $Re_2$-$S_2$ | 2.374 |
| $Re_2$-$S_3$ | 2.422 |
| $Re_2$-$S_4$ | 2.468 |

The most marked change in the unit-cell angles, with increasing pressure (table 5), concerns $\beta$: this indicates rotation of the sulfur atoms around the rhenium atoms during compression. Fitting the experimental data on the triclinic phase to the third-order Birch-Murnaghan equation-of-state yields a bulk modulus of 23GPa, with a pressure derivative of 29 while a second-order Birch-Murnaghan equation-of-state yields a bulk modulus of 49GPa.

Rhenium disulfide is a van der Waals layered semiconductor which somewhat resembles better known transition-metal dichalcogenides such as molybdenum sulfide. Its symmetry is lower however and consists only of an inversion center. Turning a layer upside-down (i.e. applying a C2 rotation about an in-plane axis) is therefore not a symmetry operation, but reverses the sign of the angle between the two non-equivalent in-plane crystallographic axes. A given layer can then be placed on a substrate in two symmetrically non-equivalent but energetically similar manners.

*Table 5 Unit-cell parameters for ReS$_2$ as a function of pressure*

| Pressure (GPa) | a(Å) | b(Å) | c(Å) | α(°) | β(°) | γ(°) | V(Å$^3$) |
|---|---|---|---|---|---|---|---|
| 0 | 6.428 | 6.508 | 6.458 | 121.21 | 87.6 | 106.77 | 54.75 |
| 1.1 | 6.320 | 6.427 | 6.450 | 120.88 | 88.7 | 107.2 | 53.03 |
| 4.4 | 6.154 | 6.363 | 6.446 | 120.72 | 91.2 | 107.61 | 50.51 |
| 6.4 | 6.101 | 6.344 | 6.444 | 120.78 | 91.9 | 106.77 | 49.54 |

This is expected to lead to a new source of domain structure in large-area layer growth. Few-layer rhenium disulfide samples have been produced - via micromechanical cleavage - which had a controlled up, or down, orientation[41]. Polarized Raman microscopy could be used to distinguish between the two orientations.

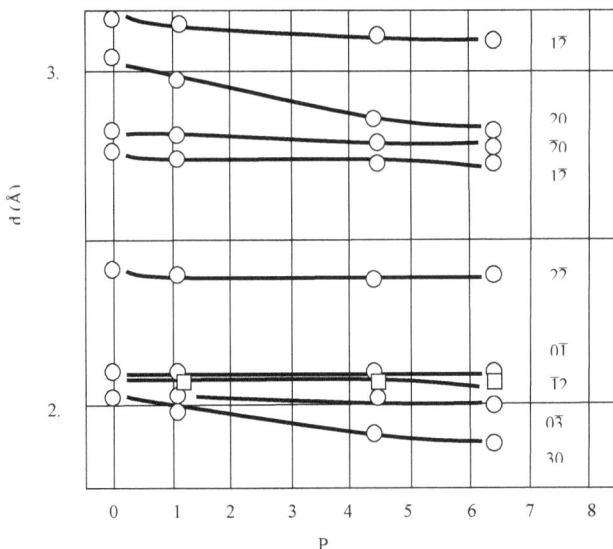

*Figure 3 Pressure dependence of the interplanar spacings of ReS$_2$*

Materials Research Forum LLC

doi: http://dx.doi.org/10.21741/9781945291920

With regard to alloy structures, a scanning transmission electron microscopic analysis of the atomic configurations of sulfur and selenium atoms in anisotropic $ReS_{1.4}Se_{0.6}$ has permitted the identification and quantification of rhenium, selenium and sulfur atoms at the various coordination sites[42]. Unlike the random distribution of metal and chalcogen elements which is found in the $MoS_{2(1-x)}Se_{2x}$ and $Mo_{1-x}W_xS_2$ analogues, the selenium atoms are preferentially located within the $Re_4$ diamonds of $ReS_{2(1-x)}Se_{2x}$. Density functional theory calculations also reveal electronic structure modulation of the selenium occupation at various sites. Experimental X-ray photo-electron spectroscopy and density-functional theory investigations[43] of $Mo_{1-x}Re_xS_2$ (x = 0.05, 0.10, 0.15 and 0.20) solid solutions showed that clustering occurs, even at low dopant concentrations of rhenium atoms. The formation of dimer-like features was observed at a rhenium concentration of 5% (table 6). An increase in the dopant concentration led to an increase in the number of clustered rhenium atoms, and to the formation of rhombic clusters. The absence of magnetism in these solid solutions suggested a mechanism for the distribution of rhenium within molybdenum disulphide via the initial formation of rhenium disulphide and its subsequent spreading.

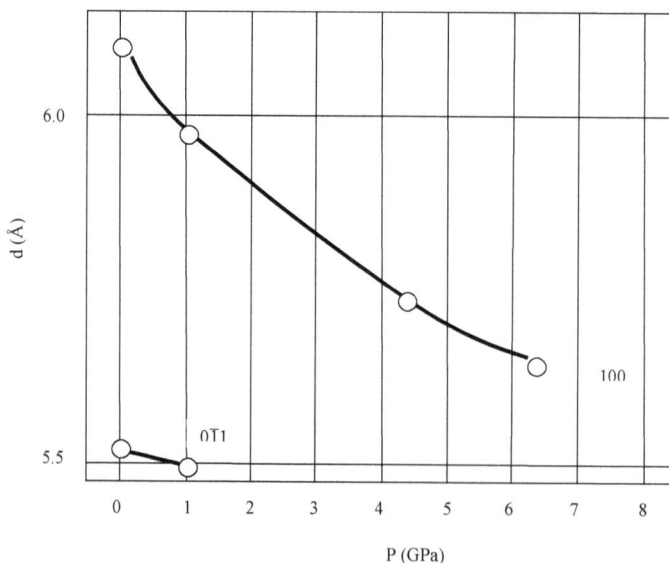

*Figure 4 Pressure dependence of the interplanar spacings of $ReS_2$*

*Table 6 Binding energies, integral intensities, spectral components and atomic environments in $Mo_{1-x}Re_xS_2$*

| x | Component | Binding Energy (eV) | Intensity (%) | Environment |
|---|---|---|---|---|
| 0 | Mo $3d_{5/2}$ | 229.6 | 100 | surrounded by 6 Mo atoms |
| 0 | S $2p_{3/2}$ | 162.4 | 96.7 | unconnected to any Re atom |
| 0.05 | Mo $3d_{5/2}$ | 229.6 | 75.4 | surrounded by 6 Mo atoms |
| 0.05 | Mo $3d_{5/2}$ | 229.1 | 24.6 | adjacent to a Re atom |
| 0.05 | S $2p_{3/2}$ | 162.5 | 69.3 | unconnected to any Re atom |
| 0.05 | S $2p_{3/2}$ | 162.0 | 22.4 | connected to a Re atom |
| 0.05 | Re $4f_{7/2}$ | 41.4 | 100 | Re-Re dimer |
| 0.10 | Mo $3d_{5/2}$ | 229.6 | 54.3 | surrounded by 6 Mo atoms |
| 0.10 | Mo $3d_{5/2}$ | 229.2 | 45.7 | adjacent to a Re atom |
| 0.10 | S $2p_{3/2}$ | 162.5 | 49.7 | unconnected to any Re atom |
| 0.10 | S $2p_{3/2}$ | 161.9 | 31.7 | connected to a Re atom |
| 0.10 | Re $4f_{7/2}$ | 41.4 | 100 | Re-Re dimer |
| 0.15 | Mo $3d_{5/2}$ | 229.6 | 55.1 | surrounded by 6 Mo atoms |
| 0.15 | Mo $3d_{5/2}$ | 229.4 | 42.2 | adjacent to a Re atom |
| 0.15 | S $2p_{3/2}$ | 162.4 | 68.1 | unconnected to any Re atom |
| 0.15 | S $2p_{3/2}$ | 161.7 | 11.9 | connected to a Re atom |
| 0.15 | Re $4f_{7/2}$ | 41.2 | 60.1 | Re-Re dimer |
| 0.15 | Re $4f_{7/2}$ | 41.6 | 9.10 | Re2-Re cluster |
| 0.20 | Mo $3d_{5/2}$ | 229.6 | 57.3 | surrounded by 6 Mo atoms |
| 0.20 | Mo $3d_{5/2}$ | 229.3 | 40.5 | adjacent to a Re atom |
| 0.20 | S $2p_{3/2}$ | 162.3 | 85.1 | unconnected to any Re atom |
| 0.20 | S $2p_{3/2}$ | 161.4 | 4.20 | connected to a Re atom |
| 0.20 | Re $4f_{7/2}$ | 41.4 | 63.8 | Re-Re dimer |
| 0.20 | Re $4f_{7/2}$ | 41.8 | 28.9 | Re2-Re cluster |
| 1.0 | S $2p_{3/2}$ | 162.6 | 54.3 | unconnected to any Re atom |
| 1.0 | S $2p_{3/2}$ | 161.9 | 36.9 | $ReS_2$ metal |
| 1.0 | Re $4f_{7/2}$ | 42.1 | 84.3 | Re-Re dimer |
| 1.0 | Re $4f_{7/2}$ | 41.5 | 12.8 | ReS2 metal |

The chemical vapor deposition and growth dynamics of highly anisotropic two-dimensional lateral heterojunctions between pseudo one-dimensional rhenium disulfide and isotropic $WS_2$ monolayers have been recently reported[44]. The disulfide and the layers have very different atomic structures: anisotropic 1T′ and isotropic 2H, respectively. High-resolution scanning transmission electron microscopy, electron energy loss spectroscopy and angle-resolved Raman spectroscopy data provide the first atomic-scale observation of the interfaces between the dissimilar two-dimensional materials. The results reveal that rhenium disulfide lateral heterojunctions with $WS_2$ produce well-oriented highly anisotropic rhenium chains which are perpendicular to $WS_2$ edges. When vertically stacked, the rhenium chains orient themselves along the $WS_2$ zig-zag direction and consequently exhibit six-fold rotation; resulting in a loss of macroscopic-scale anisotropy. The degree of the anisotropy of rhenium disulfide on WS2 depends largely upon the domain size, and decreases with increasing domain size. This is due to randomization of the rhenium chains and to the formation of disulfide sub-domains. This work established the growth dynamics of atomic junctions between novel anisotropic/isotropic two-dimensional materials, and marked the very first demonstration of control over anisotropy direction; this being a significant advance in the large-scale nano-manufacture of anisotropic systems.

The surface of the layered chalcogenide was examined using atomic force microscopy and scanning tunnelling microscopy, and the structure was determined by means of single-crystal X-ray diffraction measurements[45]. The results showed that scanning tunnelling microscopic images are associated with the surface sulfur atoms. On the other hand, bulk band-structure calculations predicted that the main contribution to the scanning tunnelling microscopic images should be due to the rhenium atoms[46]. The surface-to-tip scanning tunnelling microscopic images were also quite different to the tip-to-surface images. The surface atomic structure cannot be unambiguously deduced only from an analysis of atomic-resolution scanning tunnelling microscopic images.

Because the optical and electronic properties of the disulfide can be little modified by increasing the number of layers, stress and pressure remain the only effective means for changing the properties. This is because they directly affect the lattice parameters, symmetrise the structure and remove the Peierls distortion. Layered transition-metal dichalcogenides also exhibit novel physical properties under pressure, such as phase transitions and metallization. They are markedly sensitive to the deviatoric stress. An artificially created deviatoric stress, rather than hydrostatic pressure, can greatly affect the electronic properties of two-dimensional materials; especially anisotropic materials of the present type. When the structural, vibrational and electronic properties were investigated at up to about 34GPa by means of Raman spectroscopy, alternating-current impedance spectroscopy, atomic force microscopy and high-resolution transmission electron microscopy, together with first-principles calculations for two pressures, the experimental results showed that the disulfide undergoes a structural transformation at about 2.5GPa under both non-hydrostatic and hydrostatic conditions[47]. High-pressure Raman experiments were performed in a 300μm diamond anvil cell, using a 4:1 mixture of methanol and ethanol as the pressure medium for hydrostatic conditions. Spectra were obtained using a 514.5nm argon-ion laser at a power of less than 1mW, following a pressure standing-time of 900s. Alternating-current impedance spectroscopy was used to perform electrical conductivity measurements at frequencies of 0.1Hz to 10MHz. Metallization occurred at about 27.5GPa under non-hydrostatic conditions and at about 35.4GPa under hydrostatic conditions. The occurrence of a distinct metallization point was attributed to the influence of deviatoric stresses which markedly affected the layer structure and the weak van der Waals interactions of the disulfide. Looked at in more detail, most of the expected Raman-active modes could be seen, with prominent peaks at 150, 161 and 212/cm corresponding to the in-plane vibration modes. The Raman shift decreased with pressure, below about 2.5GPa, and then gradually increased above 2.5GPa under non-hydrostatic conditions. The marked change at this point was attributed to a phase transition from a distorted 3R structure to a distorted 1T structure; achieved by the relative sliding of sandwiched layers. A similar trend occurred under hydrostatic conditions. It was suggested that deviatoric stresses can be ignored if the pressure is not sufficiently high, given that it is known that non-hydrostatic deviatoric stresses are very small below 2.14GPa. The phase transition was also reversible, as demonstrated that the Raman spectra recovered following decompression for both the non-hydrostatic and hydrostatic situations. In impedance-spectroscopic Nyquist plots, semi-circular arcs which represented the grain boundary and grain interior resistance, both gradually increased with pressure. At above about 3GPa, the semi-circular arcs shrank and the grain-boundary resistance disappeared at about 9.5GPa. Impedance spectra were visible only in the fourth quadrant at above about 15GPa; thus indicating that the electronic crystal structure changed markedly there and might reflect a pressure-induced electronic polarization. The electrical conductivity decreased down to about 2.5GPa, and then suddenly increased thereafter under non-hydrostatic conditions. The sudden change was attributed to the effect of a phase transition, consistent with the above Raman data. The conductivity tended to remain stable at pressures of up to about 27.5GPa. The magnitude of the conductivity indicated that the disulfide was possibly metallized at above this pressure. The conductivity increased with increasing temperature at pressures of up to 12.1GPa, in a manner which was typical of semiconductor behaviour. The conductivity then fell off slightly between 12.1 and 27.5GPa, suggesting the existence of an intermediate state between semiconductor and metal. At above 27.5GPa, the conductivity decreased with increasing temperature and there was an overall variation of 12% in the conductivity between 100 and 300K; reflecting typical metallic behaviour. Given that the principal conduction mechanism in metals involves free electrons, it can be argued that with increasing temperature the scattering effect upon free electrons is increased by thermal vibration of the lattice, thus reducing the conductivity. The results of these temperature-dependent conductivity measurements thus clearly confirmed metallization of the disulfide at high pressures. Under hydrostatic conditions, the pressure required for metallization was 35.4GPa. A marked delay in metallization under hydrostatic conditions was attributed to the effect of deviatoric stresses, which have marked effect upon compression measurements. Under non-hydrostatic conditions, distortion of the geometry of the unit cell and of molecules under deviatoric stresses, leads to a lowering of the symmetry. This leads to greater interaction of the disulfide interlayers and results in observed metallization changes. The surface morphology following decompression under non-hydrostatic conditions was found to be destroyed and it was difficult to distinguish the surface layers. The surface-layer morphology was well-preserved and clearly visible following decompression under hydrostatic conditions. High-resolution transmission electron microscopic observations, performed following decompression under non-hydrostatic conditions, revealed that the layer structure was highly disordered. The layered structure was however perfectly preserved following decompression under hydrostatic conditions. These results clearly revealed the role that deviatoric stresses play in electronic-structure transitions. It was also suspected that the degree of deviatoric stress could affect the metastability of quenched materials, and that those stresses were more important in phase transitions than was the absolute magnitude of the pressure. Under non-hydrostatic conditions, the disulfide could experience much greater stresses due to a deviatoric component and stronger interlayer interaction could provoke greater structural change and a continuous lattice response; further encouraging the onset of metallization. Theoretical calculations for non-hydrostatic conditions predicted that deviatoric stress attained about 8GPa while the center pressure was about 25GPa. Calculated band-structure results revealed that the material had a theoretical direct band-gap of 1.371eV under ambient pressure. The band-gap gradually decreased with increasing pressure, and was zero at 33GPa, under non-hydrostatic conditions; thus indicating a semiconductor-to-metal transition … in line with the experimental data. The predicted density of states indicated that the highest occupied valence bands and conduction bands around the Fermi energy were dominated by the rhenium d state, plus some contribution from the sulfur p state; the two being hybridized. The degree of hybridization and electronic coupling increased with pressure, leading to diffusion of the valence and conduction bands. With increasing pressure, the high-energy valence bands widened more than did the conduction bands and this led to a decrease in the band-gap and an increase in electrical conductivity. The theoretical critical pressure for metallization was about 9GPa higher for hydrostatic conditions than for non-hydrostatic conditions. Overlap of the conduction and valence bands was assumed to provoke metallization. Reduction of the spacing between layers under high pressures is expected to decrease the band-gap of all types of layered transition-metal dichalcogenides. Overlap of the conduction and valence bands when the band-gap becomes zero then results in metallization. In the case of the present material, the band-gap could decrease with increasing compressive or tensile uniaxial strain and also decrease with increasing shear strain, thus making it more sensitive to deviatoric stress due to the anisotropic structure.

The structural properties, stability and electronic properties of single-walled $ReS_2$ nanotubes were first studied[48] using the density-functional tight-binding method, showing that the properties of such nanotubes are governed mainly by the electronic structure, which leads to an unique intralayer metal-metal bonding within the nanotube walls. This in turn explains their semiconducting nature, and a stiffness which rivals that of carbon and BN nanotubes. The stability of the nanotube is characterised by its strain energy relative to the energy of a monolayer (figure 5), and is proportional to the inverse square of the radius. This figure shows that the strain energy of the nanotube is almost independent of its chirality, except when the radius is less than 6Å. Within that range the strain energy is high enough to cause the nanotubes to disintegrate into bundles of nanostripes having dangling bonds at their edges, and so rhenium disulfide nanotubes with such small dimensions are unlikely to be observed. The strain energies of $ReS_2$ nanotubes are of the same order as those of $MoS_2$ nanotubes having the same radius. The mechanical properties in the axial direction are different however. The Young's modulus of a 12Å-radius $ReS_2$ nanotube ranges from 370 to 400GPa and is greater than that of nanotubes of other layered dichalcogenides. Rhenium disulfide nanotubes also tend to be among the most rigid inorganic nanotubes. This was attributed mainly to the presence of intralayer covalent bonding between the metal atoms of the $ReS_2$. Such body is nearly absent in other dichalcogenides. All rhenium disulphide nanotubes having radii greater than 10Å are semiconductors. The band-gap increases with increasing radius, and ranges from 0.59 to 1.11eV; the latter value being that for a monolayer. Calculations performed using the extended Huckel tight-binding, localized spherical wave and full-potential linearized augmented plane-wave methods predicted values of 1.27, 1.0 and 1.16eV, while optical absorption and high-temperature conductivity measurements yielded values of 1.32, 1.37 and 1.19eV, respectively. Nanotubes having radii of less than 10Å are metallic.

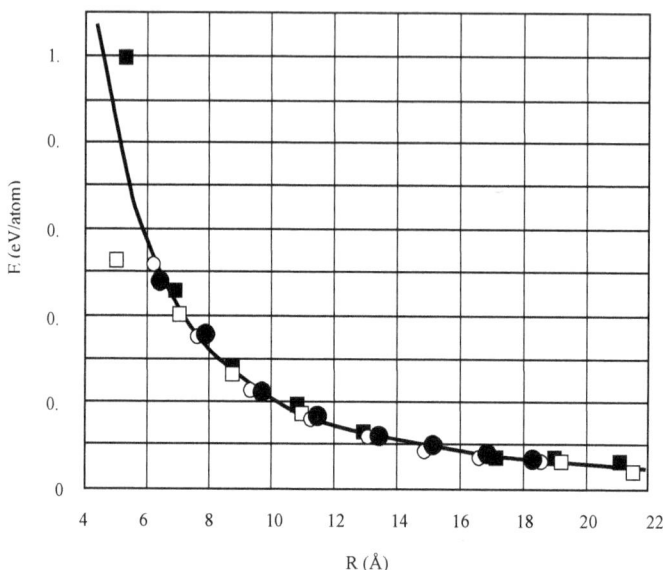

*Figure 5 Strain-energy of ReS₂ nanotubes as a function of the radius.
Open circles: (n,n\*) armchair, open squares: (n,0) zig-zag,
closed squares: (0,n) zig-zag, closed circles: (n,n) armchair*

This is attributed to the greater overlap between rhenium states. The density-of-state levels of (20,0) and (12,12) $ReS_2$ nanotubes with radii of 10.8 to 11.7Å all exhibited the same typical general features regardless of their chirality and of the ordering of the $Re_4$ clusters. Core-like sulfur 3s-states were localized at -16 to -13.5eV with respect to the Fermi level. Sulfur 3p-states, hybridized with rhenium 5d-states, formed a valence band lying between -8 and -2eV. A band between -2 and -0.5eV, and the bottom of the conduction band, were composed of rhenium 5d-states. Such an arrangement was very similar to the density-of-states of a $ReS_2$ monolayer. The shape of the density-of-states near to the Fermi level, and the semiconduction, could be easily explained in terms of a simple rigid-band model for the electronic structure of d-metal dichalcogenides. The rhenium atom has a valence-shell configuration of $d^5s^2$, and loses two s-electrons and two d-electrons. The remaining three so-called non-bonding d-electrons can redistribute into the fully-occupied $d_{z2}$-orbital and partly occupy orbitals at higher energies of $d_{x2-y2}$ and $d_{xy}$ type. In contrast to the semiconducting molybdenum and tungsten dichalcogenides with their 2 non-bonding d-electrons, rhenium disulfide should be metallic, as calculations for ReS$_2$ with hexagonal symmetry confirm. This partial filling leads to structural instability and to the formation of Re–Re bonds which open the gap and elevates the structure into the energetically more favourable triclinic form. Geometry-optimization calculations performed on zig-zag ReS$_2$ nanotubes in the hexagonal modification have confirmed the formation of rhenium-atom clusters and a transformation into nanotubes based upon the triclinic structure. Delocalization of d-electrons and the formation of Re-Re bonds is supported by an analysis of the bond orders, as calculated via Mulliken population analysis. For Re–Re bonds which form the edges of Re$_4$ clusters, they are about 0.3e and, for the one in the middle of the Re4 it is 0.36e. For Mo–Mo, the bond order is about 0.0e. The rather small charge on Re in these systems also plays a role. The Mulliken charge of the rhenium atom within a ReS$_2$ monolayer or a nanotube is about +0.3e; much smaller than the +0.9e for a molybdenum atom in a MoS$_2$ monolayer. Thus ReS$_2$ nanotubes exhibit unique features because, due to the d$^3$-configuration, the rhenium atoms form Re–Re bonds and are connected – via them - to chains of Re$_4$ units within the ReS$_2$ layers. This breaks the hexagonal symmetry of the monolayer which is typical of other d-metal dichalcogenides, and the ReS$_2$ lattice is free to take on a triclinic structure which in turn permits the existence of two sets of nanotubes having the same so-called classical chiral numbers as the nanotubes of hexagonal compounds. The formation of Re–Re bonds also imparts semiconducting properties to rhenium disulfide monolayers and nanotubes.

A significant disadvantage to the use of layered chalcogenide nanoparticles and nanotubes is their resistance to chemical and biological modification and functionalization. Improvement of chalcogenide/matrix interface bonding markedly advances the use of the former in composite materials. One method[49] for modifying layered chalcogenide nanoparticles is based upon the chalcophilic affinity of metals and the use of chelating terpyridine. The anchor group of the latter can be conjugated to fluorescent tags or to hydrophilic/hydrophobic groups. These then confer solubility in various solvents upon otherwise-insoluble chalcogenide nanoparticles.

**Defects**

First-principles calculations of monolayer rhenium disulfide containing vacancies show that the S$_4$ defect is more likely to form than is any other type of vacancy. Asymmetrical deformation caused by strain causes the band structure to transform from direct band-gap to indirect band-gap. Analysis of the partial density of states indicates that the rhenium d-orbital and sulfur p-orbital are the main components of the defect states; these being located both above and below the Fermi level[50]. The effective mass is sensitive to, and anisotropic with regard to, external strain. The reflection spectrum can be extensively tuned by external strains, suggesting that rhenium sulfide monolayers can function as nanoscale strain sensors. Such monolayers can be produced, using conventional mechanical exfoliation techniques, from as-grown rhenium disulfide crystals. Irradiation with 3MeV $He^{2+}$ ions has been used to break the strong covalent bonds in the flakes, and photoluminescence measurements show that the luminescence arising from monolayers is largely unchanged following high-energy α-particle irradiation.

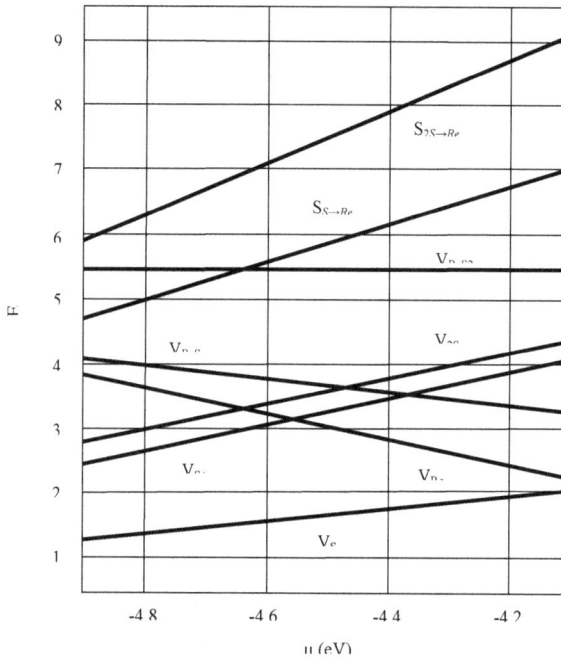

*Figure 6 Defect formation energies in monolayer ReS$_2$ as a function of the sulfur chemical potential*

In addition to the distorted 1T crystal structure, the formation of imperfections such as point defects and grain boundaries markedly affects the electronic, optical, mechanical and magnetic properties. Imperfections are created during crystal growth or exfoliation. Their existence creates localized electronic and excitonic states which can alter optical absorption, charge carrier generation, separation, and transport dynamics. The formation of sulfur and rhenium vacancies is very likely to occur, and the most common vacancies are mono-sulfur vacancies, $V_S$, di-sulfur vacancies, $V_{2S}$, double mono-sulfur vacancies, $V_{S+S}$, rhenium vacancies, $V_{Re}$, complexes of rhenium and mono-sulfur or di-sulfur, $V_{ReS}$,

$V_{Re2S}$, and sulfur-substituted rhenium vacancies, $S_{S \to Re}$, $S_{2S \to Re}$ (figure 6). The stability of these point defects depends upon the formation energy in a $ReS_2$ super-cell. Among these point defects, the mono-sulfur vacancy (figure 7) is more likely to occur because it has the lowest formation energy. The presence of sulfur vacancies could lead to non-magnetic semiconducting ground-states while rhenium vacancies could lead to spin-polarized ground-states having localized magnetic moments of 1 to $3\mu_B$. They are less likely to occur however because large amounts of energy are required to create such a vacancy. The coexistence of sulfur and rhenium point defects does not impair the semiconducting properties, apart from a slight change in the band-gap. The degree of change from 1.47, to 1.27 or 1.08eV, depends upon mono- and di-sulfur vacancies, respectively.

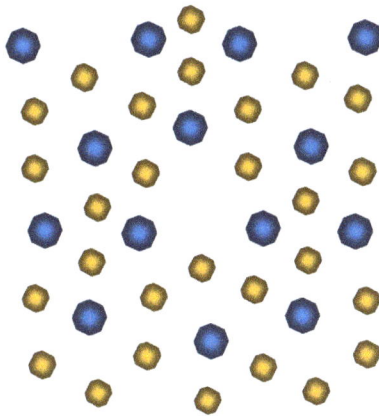

*Figure 7 Atomic structure of a monosulfur vacancy, $V_S$*

First-principles calculations reveal that the formation of a single sulfur vacancy requires the lowest formation energy under both rhenium-rich and sulfur-rich conditions, and that a random distribution of such defects is energetically preferred[51]. The sulfur point defects do not lead to any spin polarization, while the creation of Re-containing point defects induces magnetization; with a net magnetic moment of 1 to $3\mu B$. The experimentally observed easy formation of sulfur vacancies is in accord with first-principles calculations.

High-quality rhenium disulfide, prepared via chemical vapor deposition, was used to create a high-performance field effect transistor with an on/off ratio of about $10^5$. By combining electrical transport measurements at 80 to 360K, with first-principles calculations, sulfur vacancies were identified (figure 8) which exist intrinsically in rhenium disulfide and markedly affect the performance of rhenium disulfide field effect transistor devices[52]. These sulfur vacancies can efficiently adsorb and identify oxidizing gases such as oxygen and reducing gases such as ammonia, which interact electronically with rhenium disulfide only at defect sites.

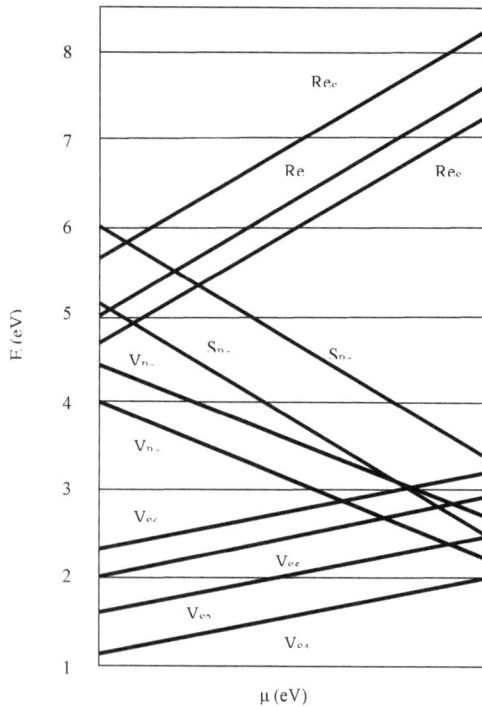

*Figure 8 Formation energy of defects in ReS₂ nanosheets
as a function of the sulfur chemical potential*

Single-layer rhenium disulfide has potential use as the anode in alkali-metal-ion batteries. First-principles calculations have been performed[53] in order to evaluate the use of pristine or defective monolayer ReS₂ as the anode in lithium- and sodium-ion cells. The calculations suggested that, due to its low-symmetry structure, there are various adsorption sites for lithium and sodium on pristine ReS₂. The adsorption of lithium and

sodium on pristine $ReS_2$ is very strong, with adsorption energies of -2.28 and -1.71eV, respectively. The presence of point defects causes an appreciably stronger binding of the alkali-metal atoms, with adsorption energies in the range of -2.98 to -3.17eV for lithium and of -2.66 to -2.92eV for sodium. A single Re-vacancy was the most strongly-binding defect for lithium adsorption, while a single sulfur vacancy was the strongest binding defect for sodium. Diffusion of the alkali atoms on pristine $ReS_2$ was anisotropic, with an energy barrier of 0.33eV for lithium and of 0.16eV for sodium. The energy barriers which were associated with the escape of lithium atoms from a double vacancy and from a single vacancy were large; being 0.60eV for the former and 0.51eV for the latter. The corresponding energy barriers for sodium atoms were 0.59 and 0.47eV, respectively. This implied a tendency to slower migration and to sluggish charging and discharging. On the other hand, the diffusion energy barrier over a rhenium single vacancy was just 0.42eV for lithium atoms and 0.28eV for sodium atoms. Single and double sulfur vacancies could reduce the diffusion rate by $10^3$ and $10^5$ times for lithium and sodium ions, respectively. It can be concluded that monolayer $ReS_2$ with a rhenium single vacancy adsorbs lithium and sodium more strongly than does pristine $ReS_2$.

The isotropic group-VI transition-metal chalcogenides exhibit easy crystal growth, with a well-defined domain architecture and grain boundaries. But rhenium disulfide, because of its anisotropic interfacial energy, has a stable dendritic growth mode and this prevents the establishment of a well-defined crystal orientation. It is therefore essential to understand the growth mechanisms in order to prepare highly crystalline monolayers having appreciable structural anisotropy. Insight is also required into how sub-domains arrange themselves into large-scale flakes, and the direction in which the rhenium chains are oriented in each sub-domain. The strong interaction and dimerization between adjacent rhenium atoms causes highly-oriented Re–Re chains to exist in the b-axis direction. When Re–Re chains come together at differing angles, clusters of vacancy-related defects appear and alter the direction and rotation of the rhenium chains around the b-axis. Triangular, hexagonal and other domains can form, depending upon the flow-rate, temperature and precursor types. A grain boundary forms when two or more domains meet. Angle-resolved Raman intensity mapping shows that the flakes of hexagonal domains are not composed of randomly-oriented Re–Re chains along the b-axis, but are composed of sub-domains. Within hexagonal domains, the sub-domains are triangular. The intensity of the 214/cm Raman peak attains its maximum when the polarization vector is almost parallel to the b-axis, and the origin of this peak is related to the in-plane vibrations of rhenium atoms. The direction of a rhenium chain is closely aligned for sub-domains which are located 180° apart. The domains form grain boundaries when they run from 30° to 210°, 90° to 270° or 150° to 330°. It is concluded that opposite triangular domains have a similar b-axis orientation. In truncated triangular flakes, the rhenium chain directions are oriented towards 90°, 330° or 210°; implying that rhenium chains are randomly oriented when close to those grain boundaries where the b-axis within each sub-domain is oriented perpendicularly to the truncated edge. The formation of grain boundaries in $ReS_2$ is unclear due to the anisotropy of the interfacial energy and to the number of differing atomic arrangements around defect sites.

## Electronic Structure

The band structures of bulk $ReS_2$ and $ReSe_2$ feature complicated interband transitions, and several close-lying band-gaps. Three-dimensional plots of the constant energy surfaces, in the Brillouin zone at energies close to the band extrema, show[54] that the valence-band maximum and conduction-band minimum need not be located at special high-symmetry points. Both materials are of indirect-gap type and it is necessary to consider the entire Brillouin zone volume when investigating the properties. A detailed comparison of the band structures of bulk $ReS_2$ and $ReSe_2$, using local density approximation density-functional theory, fully relativistic pseudopotentials and the projector augmented wave method, shows that only consideration of the band structure over the entire three-dimensional Brillouin zone volume can correctly determine the locations of the band extrema for any given level of theoretical precision. The classification of the interband optical transitions in $ReS_2$, using density-functional theory calculations is very difficult because direct and indirect transitions are close together in energy and are within the range of density-functional theory results which result from using different choices for the pseudopotential. Calculation and experiment are in better agreement in the case of bulk $ReSe_2$, which is thought to be an indirect semiconductor.

Photo-acoustic and modulated reflectance spectroscopic studies of the indirect and direct band gaps of van der Waals crystals such as dichalcogenides and monochalcogenides show that the indirect band gap can be determined by using photo-acoustic methods, while the direct band gap can be studied by using modulated reflectance spectroscopy. The latter is not sensitive to indirect optical transitions. By measuring photo-acoustic and modulated reflectance spectra, for a given compound, and by comparing them it is easy to deduce the band-gap characteristics and the energy-difference between the indirect and direct band gap. Both the indirect and direct band gaps in van der Waals crystals parallel familiar chemical trends in semiconductor compounds[55]. In the specific case of $ReS_2$ the separation between the energy gaps which was determined using photo-acoustic and modulated-reflectance methods was relatively small, the respective band-gaps being 1.37 and 1.55eV. It was noted that, for all of these compounds, the replacement of sulfur atoms by selenide atoms led to crystals having a larger lattice constant and a narrower

(indirect or direct) band-gap; plus a smaller energy difference between the band-gap types. A change in the transition metal type also led to a change in the crystallographic structure.

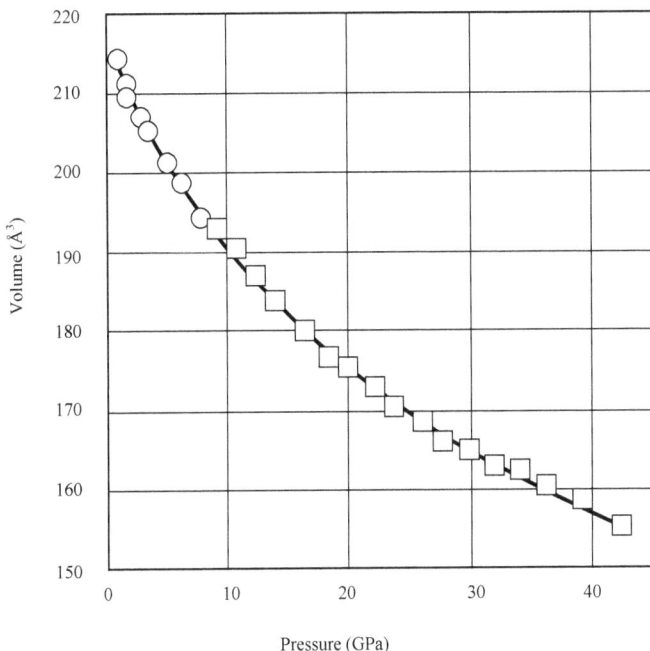

*Figure 9 Lattice volume of ReS$_2$ phases as a function of pressure*
*Circles: distorted 1T, squares: distorted 1T'*

The electrical and optical properties of ReS$_2$ under high pressures remain largely unexplored. A recent experimental and theoretical study[56] investigated its structural, electronic and vibrational properties, and visible-light response up to 50GPa (figure 9). There is a direct-to-indirect band-gap transition in the 1T phase at pressures of up to 5GPa. Under greater compression, it undergoes a structural transition to a distorted-1T' phase at 7.7GPa; followed by isostructural metallization at 38.5GPa. *In situ* Raman spectra and electronic structural analysis reveal that the interlayer sulfur-sulfur interaction is markedly increased during compression. This leads to notable changes in the electronic properties, such as band-gap closure (figure 10) and an enhanced photo-response. Pressure thus plays a critical role in fine-tuning its properties and the potential use of the layered disulfide in pressure-responsive opto-electronic applications.

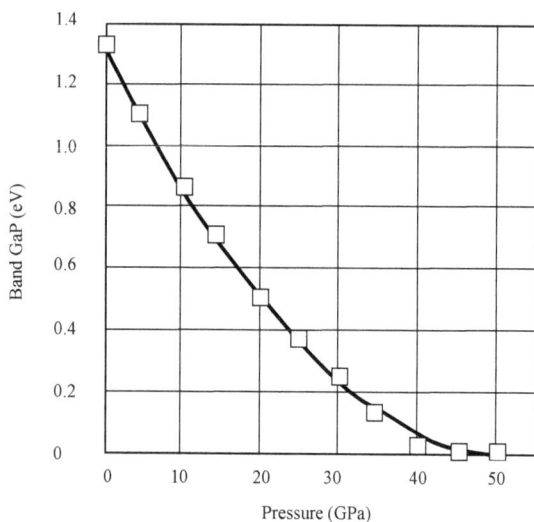

*Figure 10 Calculated band-gap of layered ReS₂ as a function of pressure*

Considering this behavior in more detail, high-pressure *in situ* powder X-ray diffraction experiments were performed using a symmetric diamond-anvil cell and a monochromatic X-ray beam having a wavelength of 0.4066Å and a diameter of 5µm. The high-pressure electrical resistance was measured using a quasi 4-probe system. A 20W incandescent lamp was used as the illumination source, with an intensity of about $2W/cm^2$ at the sample. A dark current of µA-magnitude was produced by applying a constant voltage of 0.002 to 0.3V at a given pressure. X-ray diffraction patterns showed that, under compression, all of the peaks moved to higher angles (smaller d-spacing). The (200) peak, which represented the interlayer spacing along the a-axis, changed more rapidly to smaller d-spacings and merged, with the (1$\overline{2}$0) peak, into a single broad feature at 7.7 to 18.3GPa.

The more rapid change in the (200) peak suggested that the a-axis is more compressible than are the b- and c-axes, due to interlayer de-coupling interactions. A peak appeared near to the ($12\overline{2}$) peak at 7.7GPa, and its intensity increased with increasing pressure during compression to 42.4GPa. It was deduced that a phase transition had occurred at 7.7GPa or below. The phase, distorted 1T, had a triclinic structure in which all rhenium atoms which shared the same (x,y) coordinates made up a S-Re-S triple-layer packing. In order to minimize the interlayer S---S repulsion, the angles between each 'sandwich' of distorted-1T phase were more contorted than they were in 1T-ReS$_2$. Calculations predicted that the phase transition between 1T and distorted 1T would occur at about 3.1GPa. From linear fits to plots of normalised pressure versus Eulerian strain, 8.9 to 42.4GPa, the bulk modulus was found to be 90.1GPa.

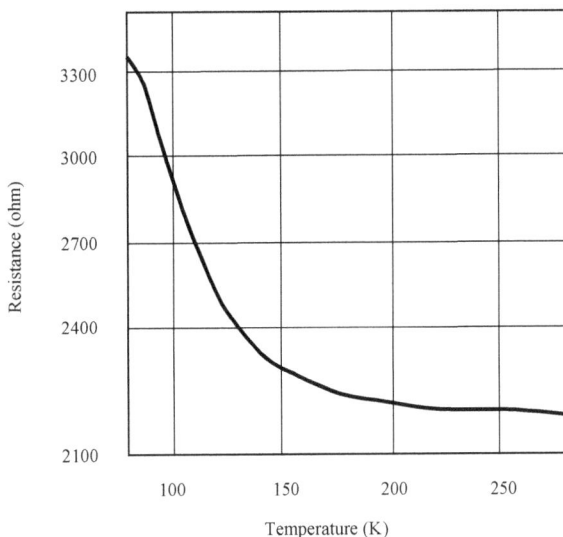

*Figure 11 Temperature dependence of the resistance of ReS$_2$ under a pressure of 12.3GPa*

The structural changes in both during compression were very anisotropic, according to the relationship between lattice parameters and pressure. The a-axis along the layering direction was more compressible than were the b- or c-axes, due to the weaker interlayer coupling force. The phase transition from 1T to 1T' was seen in *in situ* Raman data. Raman spectroscopy evidenced a large number of vibrational modes for 1T ReS$_2$, due to its low symmetry. Most Raman-active modes exhibited subtle changes under compression, plus discontinuous changes in the case of $A_g$ (438/cm) and $C_p$ (377.9/cm) at 9.4GPa. Low-frequency Raman peaks (153.1, 163.6, 217.2 and 237.2/cm) due to rhenium vibration were less sensitive to hydrostatic pressure, due to de-coupled vibrations in the $ReS_2$.

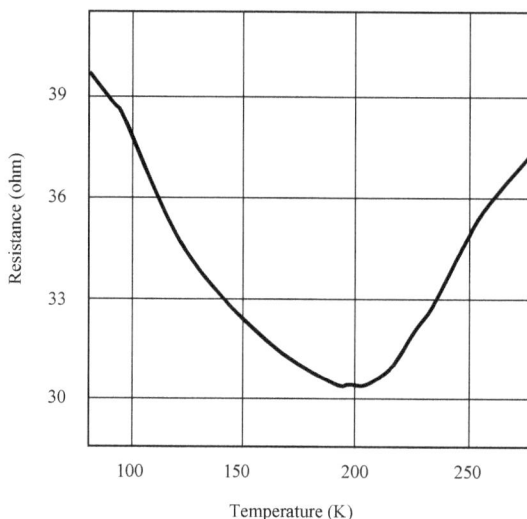

*Figure 12 Temperature dependence of the resistance*
*of $ReS_2$ under a pressure of 17.6GPa*

The pressure-dependence of the low vibrational frequencies was almost linear, indicating the occurrence of almost identical rhenium vibrations in 1T and 1T'. The slopes for $C_p$ (377.9/cm) and $A_g$ (438/cm) were much greater than those of the $E_g$ modes. The $A_g$ mode is attributed to out-of-plane sulfur vibrations at 438/cm, a surface mode, while the $C_p$ mode is just the in-plane and out-of-plane coupled mode of rhenium and sulfur atoms. Distinct features at above 250/cm arise mainly from intense sulfur-atom motion, related to layer-sliding and continuously increasing S---S interlayer interactions. The latter is also seen in the change in S-S interlayer distances, which decrease appreciably from 4.036Å at ambient pressure to 2.365Å at 65GPa. The van der Waals radii for sulfur are 1.80Å while the covalent radii are 1.05Å. The monotonic decrease suggested that the S-S interaction underwent a transition from van der Waals bonding to covalent-type. Following the formation of covalent-like S–S bonds at high pressure, the reduction in the

S-S distance begins to slow and the coupled librational and stretching motions of S–S dimers could divided into $A_g$ and Cp modes. *In situ* resistance measurements showed that the resistance fell, by 4 orders of magnitude with increasing pressure, to a stable value of less than 1ohm at 45.7GPa; suggesting the occurrence of a semiconductor to metal transition.

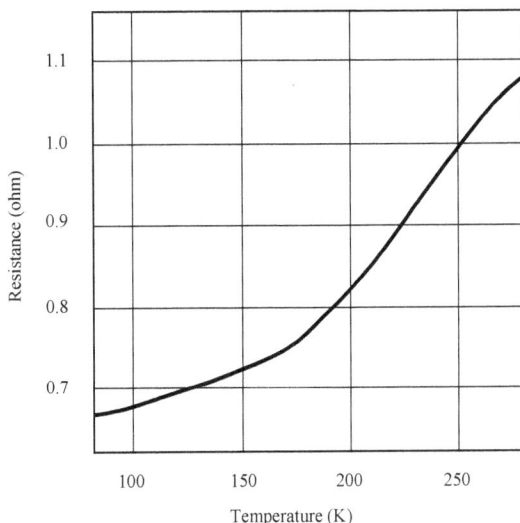

*Figure 13 Temperature dependence of the resistance of $ReS_2$ under a pressure of 38.5GPa*

*In situ* temperature-dependent resistance measurements proved the existence of metallic behavior at above 38.5GPa. The bulk disulfide exhibited typical semiconducting behavior at low pressures, and the temperature coefficient of resistance was negative at 12.3GPa (figure 11). The temperature-resistance curve at 17.6GPa (figure 12) indicated semiconduction in the low-temperature region and metallic behavior at above 200K, with a positive temperature-dependence. Complete metallization occurred at 38.5GPa (figure 13) and a positive temperature-dependence of the resistance existed at all temperatures (figure 14). *In situ* photocurrent measurements of monocrystalline and powder semiconducting samples were also carried out under high pressures (figure 15). Both types of sample exhibited a typical response to on-off switching of visible light at all pressures. With increasing pressure, the photo-response gradually increased due to continuous decreases in the resistance and band-gap. The valence electrons are therefore much more easily excited by visible light and surmount the band-gap into the conduction band. The photocurrent then greatly increases because photo-electron carriers start to migrate quickly and form a closed circuit. The bulk material could be excited at up to 39.3GPa, suggesting that the sample was a semi-metal or that semiconducting material was present due to incomplete electronic transition. The photocurrent signals from monocrystalline material disappeared at 27.8GPa; a much lower magnitude than that found for bulk material. This was attributed to an anisotropic conductivity of the monocrystalline material, due to its low lattice-symmetry.

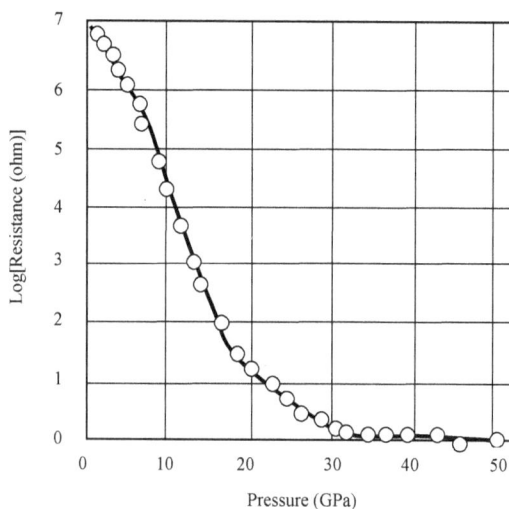

*Figure 14 Room temperature of the electrical resistance
of ReS₂ as a function of pressure*

## Properties
### Mechanical

By combining experimental data with first-principles calculations, an analysis has been made of the fracture behavior of compounds of the form, $ReX_2$. The calculations considered both monolayers and multilayers and, in the case of multilayers, the effect of the absence or presence of van der Waals forces was explored. It is noted that the cleaved edges of $ReX_2$ flakes usually form angles of about 120 or 60°.

Materials Research Forum LLC
doi: http://dx.doi.org/10.21741/9781945291920

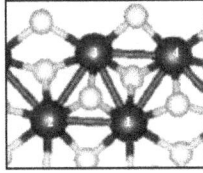

*Table 7 Ultimate tensile strengths and critical strains of ReS$_2$ in various directions*

| Direction* | Angle($^{\circ}$) | Layer | Van der Waals | UTS(GPa) | Strain |
|---|---|---|---|---|---|
| parallel to Re$_1$-Re$_2$ | 0 | mono | - | 19.69 | 0.16 |
| parallel to Re$_1$-Re$_2$ | 0 | multi | yes | 19.16 | 0.16 |
| parallel to Re$_1$-Re$_2$ | 0 | multi | no | 19.25 | 0.16 |
| perpendicular to Re$_1$-Re$_4$ | 30 | mono | - | 15.56 | 0.16 |
| perpendicular to Re$_1$-Re$_4$ | 30 | multi | yes | 15.15 | 0.16 |
| perpendicular to Re$_1$-Re$_4$ | 30 | multi | no | 15.24 | 0.15 |
| parallel to Re$_1$-Re$_3$ | 60 | mono | - | 18.41 | 0.17 |
| parallel to Re$_1$-Re$_3$ | 60 | multi | yes | 17.96 | 0.17 |
| parallel to Re$_1$-Re$_3$ | 60 | multi | no | 18.10 | 0.16 |
| perpendicular to Re$_1$-Re$_2$ | 90 | mono | - | 15.66 | 0.12 |
| perpendicular to Re$_1$-Re$_2$ | 90 | multi | yes | 15.21 | 0.12 |
| perpendicular to Re$_1$-Re$_2$ | 90 | multi | no | 15.52 | 0.11 |
| parallel to Re$_1$-Re$_4$ | 120 | mono | - | 17.51 | 0.14 |
| parallel to Re$_1$-Re$_4$ | 120 | multi | yes | 17.06 | 0.14 |
| parallel to Re$_1$-Re$_4$ | 120 | multi | no | 17.13 | 0.13 |
| perpendicular to Re$_1$-Re$_3$ | 150 | mono | - | 21.26 | 0.18 |
| perpendicular to Re$_1$-Re$_3$ | 150 | multi | yes | 20.96 | 0.17 |
| perpendicular to Re$_1$-Re$_3$ | 150 | multi | no | 21.05 | 0.17 |

*see above key

Extensive study of the uniaxial tensile stress-strain relationships of monolayer and multilayer ReX$_2$ sheets shows that the form of those particular cleaved flake edges is due

Materials Research Forum LLC
doi: http://dx.doi.org/10.21741/9781945291920

to anisotropic ultimate tensile strengths and critical strains (table 7)[57]. The calculated stress-strain relationships also explain why the cleaved edges of rhenium sulfide do not correspond to principal axes. The proposed explanation for the fracture angle is supported by the calculated cleavage and surface energies for various edge surfaces (table 8).

When the crystalline surfaces of the layered dichalcogenide semiconductor, grown via vapour-phase transport, were investigated by means of combined atomic force microscopy, lateral (friction) force microscopy, force modulation (local elasticity) microscopy and scanning electron microscopy, the as-grown crystals were found to exhibit atomically flat surfaces upon which circular islands with a typical diameter of 0.3μm and a height of 30 to 50nm were present[58]. The atomic force microscopy revealed only topographical information, but simultaneously recorded lateral force and force modulation images revealed a clear material difference. That is, on the islands the lateral forces were higher and the local elasticity was lower than on the bare rhenium disulfide surface. Analysis of the dependence of topographical and lateral force images upon the scanning direction indicated that, during crystal growth, a foreign material such as $ReBr_3$ had segregated to the surface of the disulfide crystal. Atomic force microscopy, together with lateral force microscopy and force modulation microscopy, can thus provide information concerning the composition of heterogeneous samples and the local mechanical properties of the various components.

Anisotropic two-dimensional van der Waals layered materials offer an additional dimension, as compared to isotropic two-dimensional materials, via which to tune their physical properties. Various properties of two-dimensional multi-layer materials can be modulated by varying the stacking order; due to appreciable van der Waals interlayer coupling[59].

*Table 8 Cleavage energies, cleavage strengths and surface energies for ReS₂*

| Surface | Cleavage Energy(J/m²) | Cleavage Strength(GPa) | Surface Energy(J/m²) |
|---|---|---|---|
| Containing 0° | 2.15 | 20.91 | 0.78 |
| Containing 30° | 4.06 | 29.80 | 1.20 |
| Containing 60° | 4.12 | 28.84 | 1.95 |
| Containing 90° | 3.42 | 25.46 | 1.17 |
| Containing 120° | 2.68 | 22.02 | 0.88 |
| Containing 150° | 2.88 | 23.26 | 1.06 |

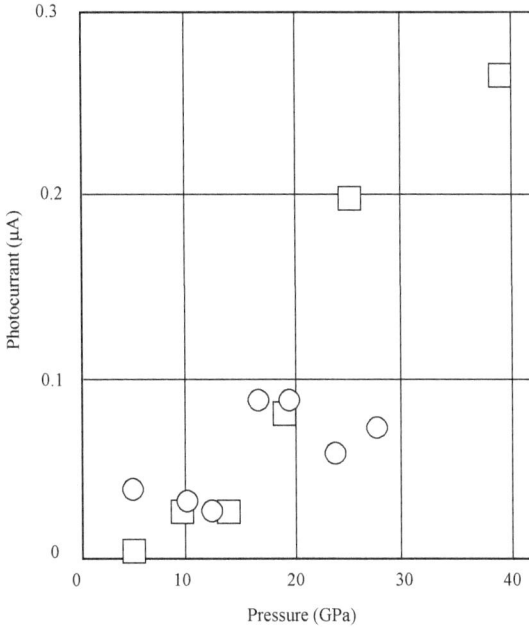

*Figure 15 Photocurrent of ReS₂ as a function of pressure Squares: powder, circles: monocrystal*

Multilayer rhenium disulfide, a typical anisotropic two-dimensional material, might be expected to be randomly stacked and to lack interlayer coupling. Two stable stacking orders, isotropic-like and anisotropic-like, were instead found using ultra-low and high-frequency Raman spectroscopy, photoluminescence and first-principles density functional theory calculations. Two interlayer shear modes are observed in anisotropic-like N-layer material (figure 16) while only one shear mode appears in isotropic-like N-layer material. This suggests the presence of anisotropic- and isotropic-like stacking orders in isotropic-like and anisotropic-like N-layer material, respectively. This explicit difference in the observed frequencies reveals an unexpectedly strong interlayer coupling in isotropic-like and anisotropic-like N-layer material. The force constants are equal to some 55 to 90% of those of multi-layer $MoS_2$.

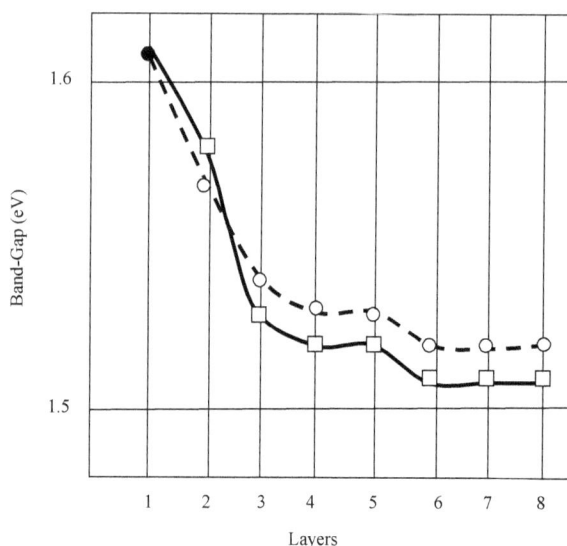

*Figure 16 Evolution of the optical transition energies of anisotropically-(circles) stacked and Isotropically (squares) stacked layers of ReS$_2$ as a function of the number of layers*

Layer-dependent Raman measurements in the ultra-low frequency range of rhenium disulfide samples with one to ten layers revealed layer breathing and shear modes, permitting easy assignment of the number of layers (figures 17 and 18). Polarization-dependent measurements also revealed an energetic shift in the shear mode which arose from the in-plane anisotropy of the shear modulus of the material[60]. An investigation was made of the interlayer phonon modes in N-layer rhenium diselenide by means of ultralow-frequency micro-Raman spectroscopy. The dichalcogenide was a distorted octahedral (1T′) phase having an appreciable in-plane anisotropy which led to sizable splitting of the in-plane layer shear modes[61]. The fan-diagrams which were associated with the measured frequencies of the interlayer shear modes and the out-of-plane interlayer breathing modes were accurately described by a finite linear chain model; permitting determination of the interlayer force constants. The latter were of the same order of magnitude as the interlayer force constants found for graphite and for trigonal prismatic transition-metal dichalcogenides such as MoS$_2$, MoSe$_2$, MoTe$_2$, WS$_2$ and WSe$_2$. This demonstrated the importance of van der Waals interactions in N-layer rhenium disulfide. The in-plane anisotropy resulted in a complex angular dependence of the intensity of all of the Raman modes. This could be exploited in order to determine the crystal orientation. The angular dependence of the Raman response also depended markedly on the incoming photon energy.

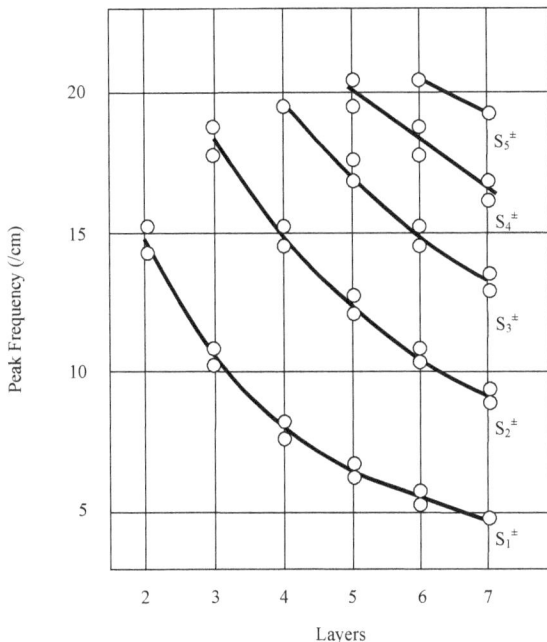

*Figure 17 Frequencies of the Raman-active interlayer shear modes of ReS₂*
*as a function of the number of layers*

The control of the optical properties and electronic structure of two-dimensional layered rhenium disulfide will be important for its use in electronic devices, but the identification of the structural transformations of monolayer disulfide produced by strain is little explored. The Raman spectra of monolayer rhenium disulfide under external strain were determined theoretically on the basis of density function theory. It is found that, due to its low structural symmetry, the deformations which are produced by external strains can control only the Raman mode intensity and cannot effect Raman mode shifts[62]. The results also suggest that the structural deformation produced by external strains can be identified by monitoring Raman scattering.

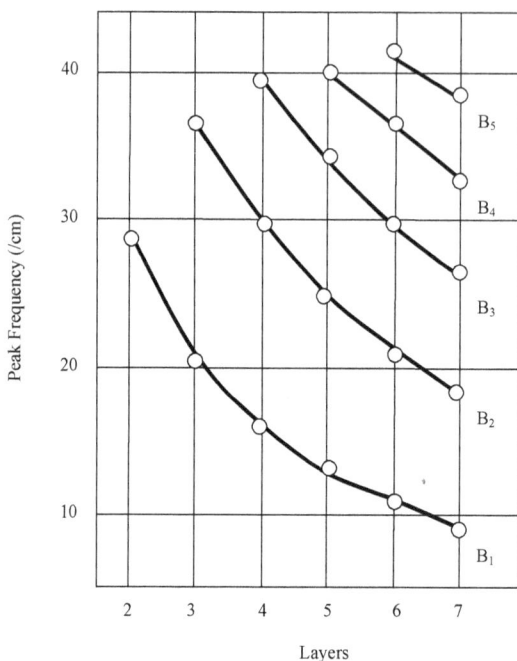

*Figure 18 Frequencies of the Raman-active interlayer breathing modes of ReS$_2$*
*as a function of the number of layers*

The class of temperature-responsive materials can adapt to the surrounding environment under the influence of a thermal stimulus, and has attracted considerable interest with regard to its use in sensors, actuators and surface engineering. The most familiar member of the class – a polymer – is however not ideal with regard to long-term reliability and durability.

Rhenium disulfide, a clearly inorganic material, nevertheless exhibits an unexpectedly marked temperature-responsive behavior which is characterized by stable and reversible thermally-induced bending. Due to thermal fluctuations in the disulfide layers, intrinsic ripples tend to amplify rapidly with increasing temperature. The already weak interlayer interaction of the disulfide is then further weakened, resulting in interlayer sliding. Due to a decrease in the bending rigidity with increasing temperature, spontaneous out-of-plane bending occurs in the rhenium disulfide layers[63].

*Table 9 Optical band-gaps of semiconducting transition-metal dichalcogenides*

| Material | Condition | Result | Band-Gap | Value (eV) |
|----------|-----------|--------|----------|------------|
| $ReS_2$ | monolayer | theoretical | direct | 1.43 |
| $ReS_2$ | monolayer | experimental | direct | 1.55 |
| $ReS_2$ | bulk | theoretical | direct | 1.35 |
| $ReS_2$ | bulk | experimental | direct | 1.47 |
| $ReSe_2$ | monolayer | theoretical | indirect | 1.34 |
| $ReSe_2$ | monolayer | theoretical | direct | 1.24 |
| $ReSe_2$ | monolayer | experimental | indirect | 1.47 |
| $ReSe_2$ | 2-layer | theoretical | direct | 1.17 |
| $ReSe_2$ | 2-layer | experimental | indirect | 1.32 |
| $ReSe_2$ | 4-layer | theoretical | direct | 1.09 |
| $ReSe_2$ | bulk | theoretical | indirect | 1.06 |
| $ReSe_2$ | bulk | experimental | indirect | 1.18 |
| $MoS_2$ | monolayer | - | direct | 1.85 |
| $MoS_2$ | bulk | - | indirect | 1.20 |
| $MoSe_2$ | monolayer | theoretical | direct | 1.34 |
| $MoSe_2$ | monolayer | experimental | direct | 1.58 |
| $MoSe_2$ | bulk | experimental | indirect | 1.10 |
| $MoSe_2$ | bulk | theoretical | indirect | 1.10 |
| $MoTe_2$ | monolayer | theoretical | direct | 1.07 |
| $MoTe_2$ | monolayer | experimental | direct | 1.10 |
| $MoTe_2$ | bulk | experimental | indirect | 1.00 |
| $MoTe_2$ | bulk | theoretical | indirect | 0.82 |

continued

*Table 9 (continued) Optical band-gaps of semiconducting transition-metal dichalcogenides*

| Material | Condition | Result | Band-Gap | Value (eV) |
|---|---|---|---|---|
| $WS_2$ | monolayer | theoretical | direct | 1.43 |
| $WS_2$ | monolayer | experimental | direct | 1.55 |
| $WS_2$ | bulk | experimental | indirect | 1.35 |
| $WSe_2$ | monolayer | theoretical | direct | 1.74 |
| $WSe_2$ | monolayer | experimental | direct | 1.65 |
| $WSe_2$ | bulk | experimental | indirect | 1.20 |
| $WTe_2$ | monolayer | theoretical | direct | 1.14 |
| $WTe_2$ | bulk | theoretical | indirect | 0.70 |

This thermally-induced bending can return to its initial state when the temperature again decreases, as confirmed by reversible wetting measurements. The thermally-induced bending of rhenium disulfide clearly offers great potential in smart-material applications.

*Optical*

Two-dimensional transition-metal dichalcogenides (table 9) possess features which make them attractive for use in the next generation of electronic and opto-electronic devices[64,65]. The appearance of the new class of two-dimensional layered materials, which possess low-symmetry crystal lattices and thus exhibit distinct electrical and optical characteristics along different in-plane crystal directions, permits the exploration of previously inaccessible domains of tunability of electrical and optical devices.[66] With the aim of obtaining tunable properties and optimal performance, great effort has thus been devoted to the exploration of two-dimensional multinary layered metal chalcogenide nanomaterials.[67]

Intensity-scan studies of polarization-dependent non-linear processes in exfoliated bulk $ReS_2$ show[68] that the absorption coefficients under high laser peak power undergo a transition from saturable absorption, to reverse saturable absorption, upon rotating the laser polarization with respect to the b-axis. Saturable absorption and excited-state absorption contribute to the non-linear optical processes, and both absorptions exhibit a marked dependence upon the polarization angle. This behaviour can be attributed to the

Materials Research Forum LLC
doi: http://dx.doi.org/10.21741/9781945291920

anisotropic optical transition probability and electronic band structure of $ReS_2$, and makes the disulfide a potential candidate as a saturable absorber for lasers and optical modulators.

The anisotropic exciton behavior of two-dimensional materials, which is induced by spin-orbit coupling or anisotropic spatial confinement, has been exploited for imaging applications. High-energy and robust anisotropic excitons have been generated[69] in few-layer rhenium disulfide nanosheets by means of phase engineering. This overcame limitations imposed by layer-thickness and permitted the production of visible polarized photoluminescence at room temperature. Ultrasonic chemical exfoliation was used to introduce the metallic T-phase of the disulfide into few-layer semiconducting $T_d$-phase nanosheets. By using this configuration, light excitation could readily produce so-called hot electrons which could tunnel to the $T_d$ phase via the metal/semiconductor interface and increase the overlap between the wave functions and screened Coulomb interactions. A marked increase in the optical band gap was observed, due to a strong electron-hole interaction. Highly anisotropic and tightly bound excitons with visible light emission (1.5 to 2.25 eV) were produced, and controlled by tailoring the T-phase concentration.

Optical absorption, photo-acoustic spectroscopy and photoconductivity were investigated in $ReS_2$ monocrystals at 50 to 300K[70]. The energy gap was 1.55eV at 80K, the average phonon energy was 17meV and the electron phonon coupling parameter was 2.40. The disulfide single crystals were grown directly from the elements in closed quartz tubes by means of vapor phase transport, using bromine or iodine as the transport agent. This produced n-type or p-type material, respectively. The quartz tubes were loaded with stoichiometric amounts of rhenium, sulfur and transport. These were pre-reacted for nearly 100h at 470C in the cold of a furnace and at up to 1000C in the hot part of the furnace. Single crystal growth took place in the same ampoule during a second stage. Crystals with areas of up to $1.532cm^2$ were obtained, having thicknesses in the mm range. The as-grown crystals had mirror-like faces which corresponded to the van der Waals planes. The lattice constant in the c-direction was found to be 6.407Å. It was puzzling that a change in conductivity-type occurred, depending upon which transport agent was used, even though the agents were in the same valence state. It was concluded that, at least for iodine, the usual concept of impurity doping by the transport agent did not apply. The n-type behavior could be explained by replacing sulfur with bromine. The explanation was more complicated for p-type samples, and one proposal was that the crystals were slightly metal-deficient; thus leading to hole conductivity. A lack of metal atoms could in turn occur because the iodides of the transition metals are not very stable, thus leading to reduced transport. Another proposal was that p-type behavior resulted from compensation of the metal-induced acceptor states and donors due to the replacement of sulfur atoms by iodine. The very different atomic radii of the iodine and the sulfur atoms made such a substitution less probable however. Energy-dispersive X-ray analysis revealed no deviation from stoichiometry, nor any evidence for incorporation of the transport agent. Because of the plate-like habit of the crystals it was possible to measure the optical absorption of samples between 8 and 80mm in thickness without special treatment. This disulfide crystallizes in a distorted $CdCl_2$ structure, leading to a triclinic symmetry in which the optical properties depended upon the polarization state of the light; even at normal incidence to the van der Waals surface. This polarization dependence was not observed near to the absorption edge. Photoconductivity measurements were performed at 400 to 1200nm and 80 to 300K, using a 300W tungsten filament lamp and a grating monochromator. During photo-acoustic measurements, the samples were illuminated with chopped monochromatic light from a Xenon arc-lamp. The advantage of photo-acoustic measurements is that scattered and reflected light from the sample surface are unimportant and sample preparation is therefore much easier. A marked light absorption at above $10^4$/cm between energies of 1.45 and 1.48eV at 300K and between 1.53 and 1.55eV at 80K was attributed to direct allowed transitions between parabolic bands. Extrapolation of the linear part of the spectral dependence permitted the determination of the value of the direct band-gap width. A value of 1.55eV at about 80K and of 1.47eV at 300K was deduced. A high photosensitivity was observed over rather a wide spectral range. The temperature dependence of the direct band-gap in n-type and p-type material was estimated from photo-acoustic spectroscopic measurements. A value of 17meV for the phonon energy led to a Debye temperature of 197K. Rhenium disulfide decomposes, without melting, at 1107K. The value of the electron–phonon coupling parameter led to the conclusion that there is a strong interaction between band-edge states and the phonon system in the disulfide. Both the n-type and p-type materials exhibit extrinsic conductivity at up to 300K. The Hall mobility is 10 to 20cm$^2$/Vs at 300K for both electrons and holes, increases with decreasing temperature down to 250 to 200K due to lattice scattering and decreases at lower temperatures due probably to scattering by ionized impurities. The maximum of Hall mobility in p-type material shifted to higher temperatures than in n-type material. This shift was attributed to a greater ionized impurity scattering in the p-type disulfide. At room temperature the electron-hole mobility was about 35cm$^2$/Vs; in good agreement with the data on Hall mobility.

Two-dimensional transition-metal dichalcogenides are active platforms for surface-enhanced Raman spectroscopy. Phase transitions lead to altered electronic structures in these materials, and a consequent effect upon Raman enhancement. By using thermally evaporated copper phthalocyanine, rhodamine 6G and crystal violet as probe molecules, it was found[71] that a phase transition from 2H- to 1T could markedly increase the Raman enhancement effect upon rhenium sulfide or selenide monolayers via a mainly chemical mechanism.

Strong and anisotropic third-harmonic generation has been detected[72] in monolayer and multilayer rhenium disulfide; the third-order non-linear optical susceptibility of the monolayer material being of the order of $10^{-18}m^2/V^2$. This is about an order of magnitude higher than the values reported for hexagonal transition-metal dichalcogenides such as $MoS_2$. A similar value for the third-order non-linear optical susceptibility was found for multilayer samples. In both monolayers and multilayers the intensity of the third-harmonic generation field depended upon the direction of the incident-light polarization. Point group symmetry analysis showed that anisotropy was not to be expected of a perfect 1T lattice and had instead to arise from the distortions of the lattice. Third-harmonic generation measurements could thus be used to characterize the lattice distortions of two-dimensional materials.

A theoretical study of the effect of dimensional reduction upon the electronic and optical properties of layered $ReX_2$ (X = S, Se) compounds considered[73] the characteristics of the band-gap under the influence of interlayer coupling. The method used was that of the self-energy corrected GW approximation (where G is the Green's function and W is the screened Coulomb interaction) as applied to optimized experimental sets of structural data. Resultant changes in the optical properties and anisotropy were monitored using the optical spectra, as obtained by solving the Bethe-Salpeter equation. At the $G_0W_0$ level of the theory, a decrease in the thickness of the sulfide from bulk to bilayer to free-standing monolayer left the band-gap direct, in spite of a change in the nature of the band-gap, with the values increasing from 1.6 via 2.0 to 2.4eV, respectively. In the case of the selenide the fundamental band-gap changes from direct, for the bulk phase (1.38eV), to indirect for the bilayer (1.73eV). It again becomes direct for a single layer (2.05eV). The results were interpreted in terms of renormalization of the band-structure.

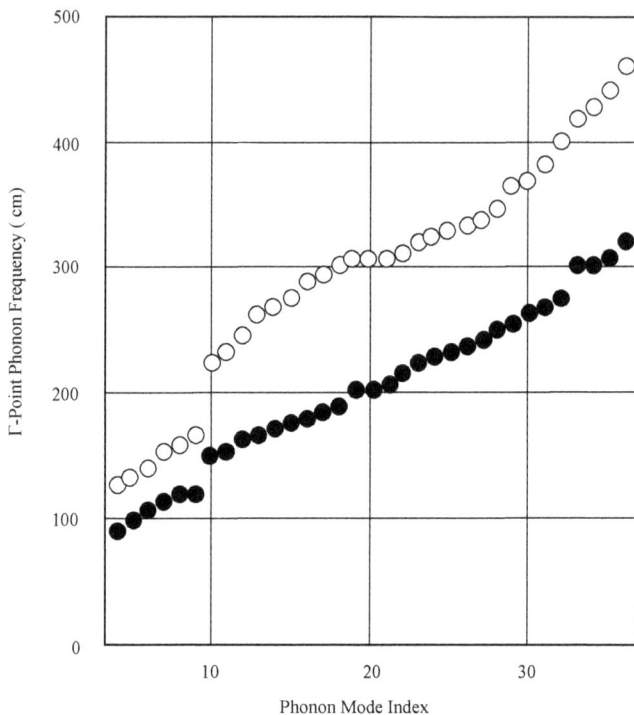

*Figure 19 Zone-centre phonon modes of ReS$_2$ (open)
and of ReSe$_2$ (closed) and related frequency*

Because the rhenium dichalcogenides are layered van der Waals semiconductors which exhibit a large number of Raman-active zone-centre phonon modes due to their unusually large unit cells and departure from hexagonal symmetry it is therefore feasible to introduce an in-plane anisotropy into composite heterostructures which are based upon such van der Waals materials and Raman spectroscopy can be used to determine their in-plane orientation. First-principles calculations can furnish a good description of the lattice dynamics of these materials and can predict the zone-centre phonon frequencies and Raman activities (table 10). The frequency distribution of the phonon modes, and their atomic displacements, offer a picture of the phonon frequencies and Raman spectra in terms of the scaling of Raman frequency with chalcogen mass[74].

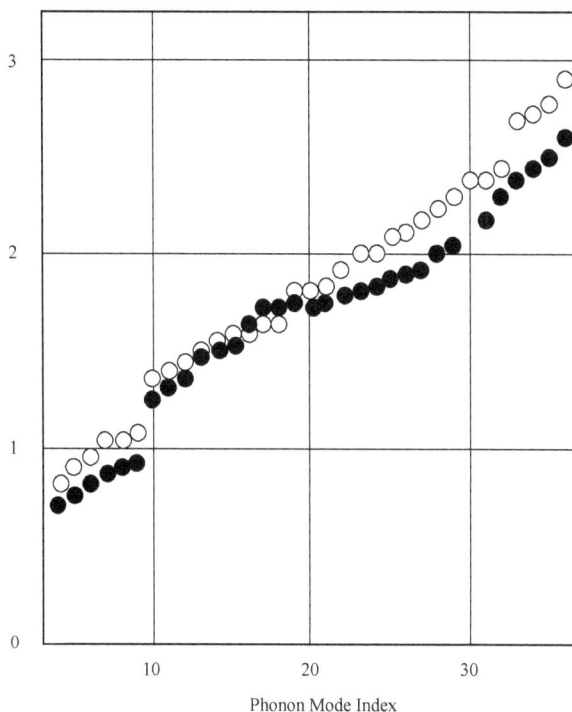

*Figure 20 The data of figure 19 as re-scaled (arbitrary units) using the square root of the chalcogen mass*

A simple argument based upon scaling of the chalcogen mass in fact leads to a unified understanding of the zone-center phonon modes of rhenium and other dichalcogenides and of the frequency distribution of the modes by exploiting the great similarity of the members of this group of materials. It is based upon a simple classification of the modes into 2 groups: those in which the displacements of the metal atoms (mainly below the gap) are significant and those in which it is mainly the chalcogen atoms (mainly above the gap) that are moving. The displacements of the metal atoms are approximately in-plane, and are appreciable for lower-frequency modes (8 or 9) but are smaller for the higher modes. In the case of those modes (10 or 11) there are appreciable out-of-plane displacements of the chalcogen atoms. Similar patterns of behavior are found for a range of compounds of this type. Considering only displacements of the chalcogen atoms, an in-plane behavior ($E_{2g}$-like) is found for mode 9 and the out-of-plane behavior ($A_{1g}$-like) for mode 10. A further elementary proposal is that, if the interatomic force constants are comparable, then the frequencies of the corresponding displacement patterns will scale directly as the square root of the force constant and inversely as the square root of the mass of the displaced atoms. For modes which are above the gap, the latter mass is that of the chalcogen. By using that simple relationship, the rather different data-points for 2 common rhenium dichalcogenides can be caused to fall on essentially the same curve (figures 19 to 21) and both sets of phonon modes almost coincide above the gap provided that the appropriate mass, metal or chalcogenide, is used.

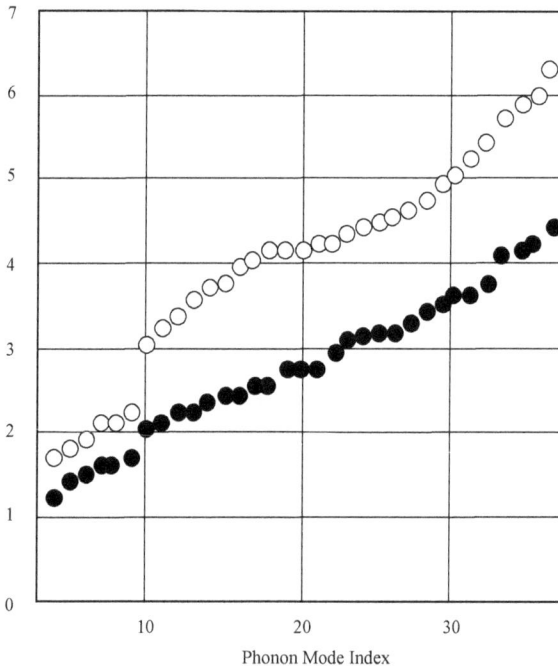

*Figure 21 The data of figure 19 as re-scaled (arbitrary units)*
*using the square root of the metal mass*

*Table 10 Calculated zone-centre phonon frequencies for ReS$_2$*

| Mode | Calculated (/cm) | Experimental (/cm) |
|---|---|---|
| 8 or 9 | 162.0 | 161.3 |
| 10 or 11 | 218.2 | 211.4 |
| highest A$_g$ | 440.7 | 438.5 |

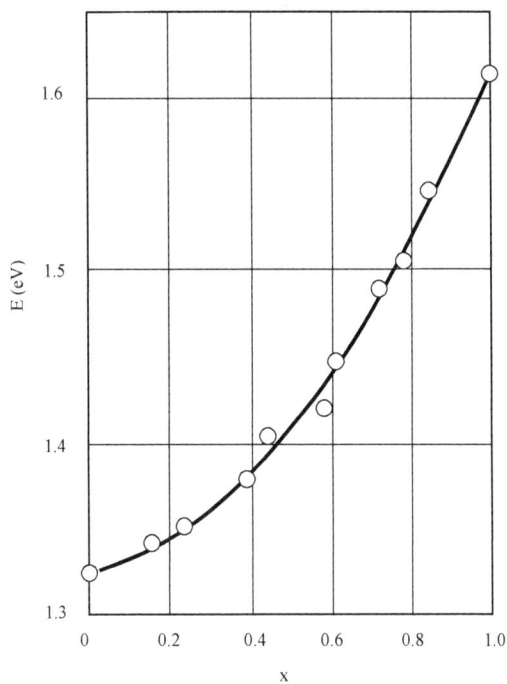

*Figure 22 Fluorescence-emission energy of ReS$_{2x}$Se$_{2(1-x)}$*
*as a function of the sulfur composition*

The chemical vapor deposition of 1T′ $ReS_{2x}Se_{2(1-x)}$ monolayers produced compositions in which the corresponding band-gaps could be continuously tuned (figure 22) from 1.32eV for $ReSe_2$ to 1.62eV for $ReS_2$ by carefully controlling the growth conditions[75]. Atomic-resolution scanning transmission electron microscopy revealed a notable local atomic distribution in the alloys, in which sulfur and selenium atoms were selectively located at various X-sites in each Re-$X_6$ octahedral unit cell, with a perfect match between the atomic radius and the volume of each X-site. Such a structure was very advantageous as it could generate any required highly-localized electronic state in the two-dimensional surface. The carrier-type, threshold voltage and carrier mobility of field effect alloy transistors could be systematically modulated by adjusting the composition. The conductivity of $ReS_{2x}Se_{2(1-x)}$ could be varied from n-type to bipolar and p-type. Due to their 1T′ structure, the compositions exhibited markedly anisotropic optical, electrical and photo-electric properties.

Anisotropic two-dimensional materials such as rhenium disulfide possess intrinsically in-plane anisotropic properties that can be tuned for use as thin-film polarizers or polarization-sensitive photo-detectors. In order to exploit these properties the crystal orientations have to be precisely determined but current means for measuring crystalline orientation, such as reflection and transmission spectra and polarized Raman-related techniques, are unreliable because of interference from sample thickness and background signals; particularly in the case of transparent substrates. Photothermal detection[76] is a new high-accuracy background-free method for the determination of the crystalline orientation of $ReS_2$ on transparent substrates. It is an optical refractive index sensing technique in which a modulated pump beam is absorbed by the disulfide, leading to a local change in the refractive index of the photothermal medium. Propagation of the probe beam at various wavelengths is modified by the periodic change in refractive index which is produced. Photothermal detection can be used to determine accurately the crystal orientation by taking account of the signal generated by anisotropic optical absorption of the pump beam and of the resultant heating of the medium. The technique suffered no background interference, regardless of the excitation wavelength or sample thickness. By studying the relationship between the polarization angle of the pump beam, and the strength of the photothermal signal, the crystal orientation can be found accurately. Detailed tests were performed on samples with a nominal thickness of 5.8nm and, because they could be easily damaged by the laser beam used, its power was restricted to 1mW. Among the characteristic Raman modes could be distinguished only the peak position of 150 and 160/cm; the other Raman modes could not be identified due to the signal being too weak. This in fact is an inherent drawback of the Raman technique, which suffers from a weak through-put for samples on transparent substrates.

Photothermal signal measurements as a function of the polarization angle of the pump beam for various thicknesses (7.5, 12.4, 20.6nm) suggested that, for a given polarization angle, the intensity of the signal would become stronger as the thicknesses of the sample was increased, due to increasing absorption of the pump beam. For a given sample, the intensity of the photothermal signal changed with the polarization angle of the pump beam due to the in-plane optical anisotropy of $ReS_2$. The polarization angles of the maximum and minimum signals thus differ by 90°. The photothermal signal had a 180° variation period. In order to confirm the worth of the photothermal detection method it was applied to the determination of the crystal orientation. The typical Raman modes (150, 213/cm) were found, and the Raman signal strength on a $SiO_2/Si$ substrate was higher than that for a $SiO_2$ substrate. This was due to an interlayer interference which also gave rise to problems in determining the crystal orientation of $ReS_2$ on a transparent substrate. The 213/cm Raman mode of $ReS_2$ can be used to the crystal's orientation because the polarization angle of the incident beam which corresponded to the maximum of the latter mode is also the direction of the rhenium chain. Photothermal detection was used to identify the crystalline orientation of samples with various thicknesses as well as using the Raman technique. It was concluded that, in the photothermal detection technique, the photothermal signal could be used as a guide in that the largest value lay along the armchair direction and smallest one lay along the zig-zag direction; regardless of the excitation wavelength and the sample thickness. It was concluded that it is difficult to apply reflectance and transmission spectroscopic techniques to the determination of the crystalline orientation of anisotropic two-dimensional materials.

Optical anisotropy is one of the most fundamental characteristics of the new two-dimensional low-symmetry materials and provides extensive structural information. An azimuth-resolved microscopic approach to direct resolution of the normalized optical difference along two orthogonal directions at normal incidence is useful because such a differential method ensures that it is sensitive only to anisotropic samples and is unaffected by isotropic materials. Experimental results[77] suggested that optical anisotropy is a suitable probe for measuring specimen thicknesses to monolayer resolution.

The optical absorption response of rhenium disulphide is such that its monolayer and bulk forms have almost identical band structures. When the non-linear saturable and polarization-induced absorption is investigated in the near-infrared communication band beyond the band-gap, a rhenium disulfide-coated D-shaped fiber exhibits a remarkable polarization-induced absorption[78]. This indicates the occurrence of differing responses with respect to transverse electric and transverse magnetic polarizations relative to the rhenium disulfide plane. The non-linear saturable absorption of the D-shaped fibers exhibits a similar saturable fluence of some tens of $\mu J/cm^2$, and a modulation depth of about 1% for ultra-fast pulses with two orthogonal polarizations. D-shaped fibers can be used as saturable absorbers to obtain self-starting mode-locking in an erbium-doped fiber laser. These features widen the operational wavelength of rhenium disulfide, from visible light to about 1550nm. Many applications could exploit the anisotropic and non-linear absorption characteristics, such as in-line optical polarizers, high-power pulsed lasers and optical communications.

Study of the anisotropic optical and transport properties of monolayer rhenium disulfide, prepared by mechanical exfoliation and subjected to transient absorption measurements using various polarization configurations and sample orientations, shows that the absorption coefficient and transient absorption are both anisotropic[79]. The maximum and minimum values occur when the light polarization is parallel to, and perpendicular to, the rhenium atomic chains, respectively. The maximum values are about a factor of 2.5 times the minimum values. By resolving the spatiotemporal dynamics of the excitons, it is found that the diffusion coefficient of excitons moving along rhenium atomic chains is about 16cm$^2$/s at room temperature. This is a factor of about three times greater than that for those moving perpendicularly to that direction. An exciton lifetime of 40ps was also deduced. The transient absorption measurements confirm that monolayer rhenium disulfide as an anisotropic two-dimensional semiconductor. Both the linear absorption and transient non-linear absorptions are a few times greater when the light polarization occurs along the rhenium atomic chain direction. The exciton diffusion is also fast along the rhenium atomic chains.

The direct energy gap, $E_g^{dir}$ = 1.55eV (80K), of rhenium disulfide and its temperature dependence were first determined by measuring optical absorption and photoconductivity in single crystals at 80 to 300K[80]. On the basis of the shape of the absorption spectrum in the indirect optical transition region ($E_g^{ind}$ = 1.27eV, 300K) the optical phonon energy was estimated to be 0.06eV.

Birefringence is an inherent optical property of materials which is due to their anisotropic crystal structures. It permits the manipulation of light-propagation parameters such as phase-velocity, reflection and refraction for use in various photonic and opto-electronic devices including wave-plates and liquid crystal displays. In this regard, two-dimensional layered materials of high anisotropy, such as rhenium disulfide, are currently of increasing interest for use in polarization-integrated nanodevice applications. Comparison of the optical birefringence of the anisotropic two-dimensional layered materials, black phosphorus, rhenium disulfide and rhenium diselenide shows that, at 520nm, the birefringence of black phosphorus (about 0.245) is some six times greater than that of rhenium disulfide (about 0.037)[81] or rhenium diselenide (about 0.047) and is comparable to that of current state-of-the-art bulk materials such as $CaCO_3$. The two-dimensional materials have been used to prepare atomically thin optical wave-plates. In the case of black phosphorus, there is a polarization-plane rotation of about 0.05° per atomic layer at 520nm. The relatively large birefringence of anisotropic two-dimensional layered materials can thus permit accurate manipulation of light polarization using an atomically controlled device thickness.

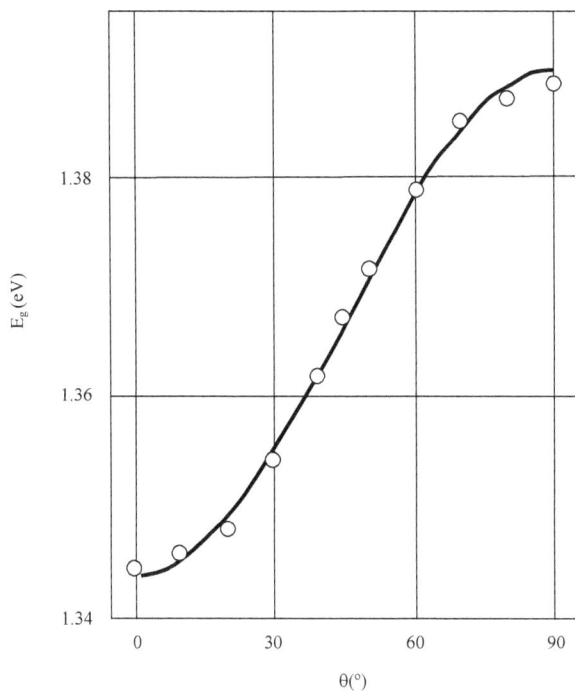

*Figure 23 Angular dependence of the polarized energy gap of ReS₂*

Optical spectroscopy also demonstrates that the reduced symmetry of rhenium disulfide, as compared to those of molybdenum and tungsten dichalcogenides, leads to anisotropic optical properties that persist from the bulk down to a monolayer. The direct optical gap blue-shifts from 1.47eV in the bulk to 1.61eV in a monolayer[82]. In the ultra-thin limit, polarization-dependent absorption and polarized emission arise from the band-edge optical transitions. Ultra-thin rhenium disulfide is thus a birefringent material with strongly polarized direct optical transitions that vary in energy and orientation as a function of sample thickness. The spectral quantum efficiency of rhenium disulfide heterodiodes was measured at wavelengths ranging from 600 to 850nm[83]. Anisotropy effects in the van der Waals plane were detected by directing linearly polarized light at normal incidence to the (001) plane of single crystals. The anisotropy of the quantum efficiency attained a maximum at 627nm. Rhenium disulfide devices are therefore well-suited to detect the polarization angle in experiments involving a HeNe laser. Due to various optical transitions, the sign of the polarization quantum efficiency changes at wavelengths greater than 750nm.

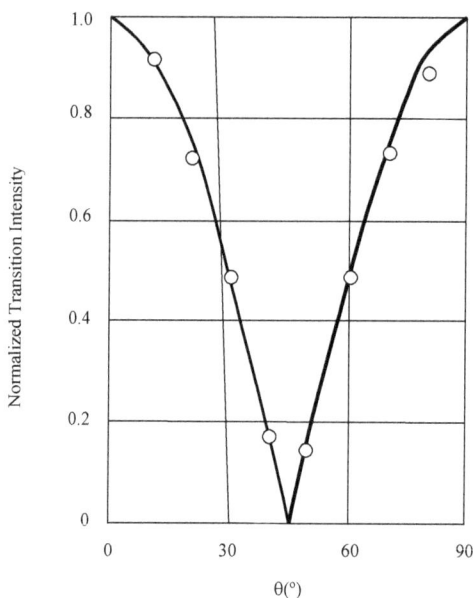

*Figure 24 Angular dependence of the normalized transition intensities of ReS$_2$*

Large monocrystals were grown directly from the component elements by means of vapor phase transport, and could be doped p-type or n-type using bromine or iodine, respectively, as the transport medium: iodine in the present case. Rhenium powder and sulfur platelets were used as starting materials. A quartz tube was loaded with 3g of

Materials Research Forum LLC
doi: http://dx.doi.org/10.21741/9781945291920

rhenium, plus a stoichiometric amount of sulfur, and placed in a 2-zone furnace at 470C: one side being held at that temperature while the other was heated to 1000C at a rate of 15C/h. These temperatures were maintained for 4 to 5 days. The reacted material was moved to one side of the transport tube and again placed in a 2-zone furnace, this time at between 1130 and 1160C; that gradient being maintained for 200h. Crystals with sizes of up to 1.5cm x 2cm and few microns in thickness were produced and the surfaces had a rod-like appearance. They were electrically characterized by making temperature-dependent Hall measurements in the van der Pauw geometry. The p-type crystals had a net carrier concentration of 7 x $10^{17}$/cm$^3$, a Hall mobility of 20cm$^2$/Vs and a resistivity of 2ohm-cm at room temperature. An activation energy of 70meV was deduced from the temperature-dependence of the carrier concentration. Photosensitive ReS$_2$ heterojunctions having an area of 3mm$^2$ were prepared by radio-frequency magnetron sputtering.

*Table 11 Energy positions of features in electrolyte-electroreflectance spectra, and inter-band transitions, of ReS$_2$*

| Feature | Energy (eV) | Assignment |
|---------|-------------|------------|
| A | 1.486 | Re 5d $t_{2g}$ → Re 5d $t_{2g}$* |
| B | 1.534 | Re 5d $t_{2g}$ → Re 5d $t_{2g}$* |
| C | 1.621 | Re 5d $t_{2g}$ → Re 5d $t_{2g}$* |
| D | 1.988 | Re 5d $t_{2g}$ → Re 5d/S 3p bonding → Re 5d/S 3p antibonding |
| E | 2.249 | Re 5d $t_{2g}$ → Re 5d/S 3p bonding → Re 5d/S 3p antibonding |
| F | 2.474 | Re 5d $t_{2g}$ → Re 5d/S 3p bonding → Re 5d/S 3p antibonding * |
| G | 2.684 | Re 5d $t_{2g}$ → Re 5d/S 3p bonding → Re 5d/S 3p antibonding |
| H | 2.952 | Re 5d $t_{2g}$ → Re 5d/S 3p bonding → Re 5d/S 3p antibonding |
| I | 3.241 | Re 5d $t_{2g}$ → Re 5d/S 3p bonding → Re 5d/S 3p antibonding |
| J | 3.716 | Re 5d $t_{2g}$ → Re 5d/S 3p bonding → Re 5d/S 3p antibonding |
| K | 4.166 | Re 5d $t_{2g}$ → Re 5d/S 3p bonding → Re 5d/S 3p antibonding |
| L | 4.486 | Re 5d $t_{2g}$ → Re 5d/S 3p bonding → Re 5d/S 3p antibonding |
| M | 4.830 | Re 5d $t_{2g}$ → Re 5d/S 3p bonding → Re 5d/S 3p antibonding |
| N | 5.344 | Re 5d $t_{2g}$ → Re 5d/S 3p bonding → Re 5d/S 3p antibonding |

Materials Research Forum LLC
doi: http://dx.doi.org/10.21741/9781945291920

Tungsten layers were first sputtered onto one side of the plate-like crystals, to serve as an ohmic back-contact, followed by heat-treatment (200C, 0.5h) in argon. Heterojunctions were then prepared by similarly magnetron-sputtering a 72nm-thick layer of n-type $In_2O_3$-5%$SnO_2$ onto the opposite side of the crystals; the resistivity of the latter layers being about 9 x $10^{-4}$ohm-cm. As a final step, electrical contacts to the emitter and tungsten back-contact were made using gold wires and silver paste. The polarization dependence of the spectral quantum efficiency was determined between 600 and 850nm, and the largest difference was observed between the quantum efficiencies at 0 and 90°; corresponding to direction perpendicular to and parallel to [010], respectively. For wavelengths greater than 770nm, the maximum in quantum efficiency occurred for a direction perpendicular to [010] and the minimum occurred for a direction parallel to [010]. At longer wavelengths, a polarization dependence was almost non-existent. A more marked structure which was found in the spectra, for directions parallel to [010], between 730 and 830nm indicated the existence of an optical transition.

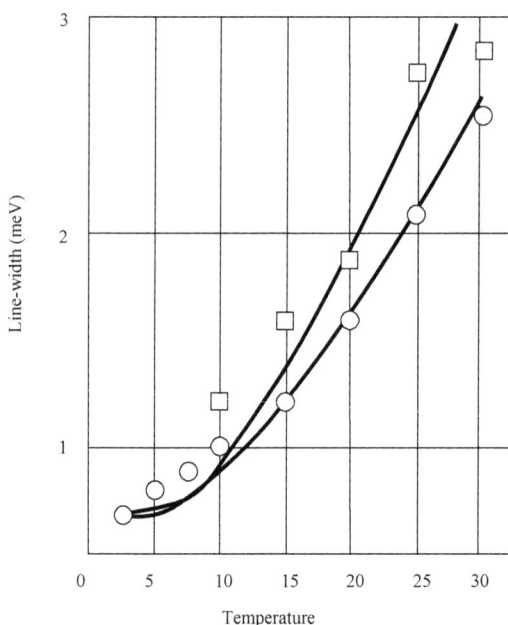

*Figure 25 Temperature dependence of the line-width of excitonic transitions in tungsten-doped ReS$_2$ Squares: $E_1^{ex}$, circles: $E_2^{ex}$*

The polarization-sensitive behaviour of band-edge transitions in the rhenium disulfide layered compound was studied by performing polarized-transmission and polarized-thermo-reflectance measurements (figures 23 and 24) at polarization angles ranging from $\theta = 0°$ (E‖b-axis) to $\theta = 90°$ (E⊥b-axis) at 300K[84]. The polarization-dependence of the polarized energy gap exhibited a sinusoidal variation with respect to the angular change in the linearly polarized light.

Polarization-dependent electrolyte-electro-reflectance measurements of rhenium disulfide layered crystals have also been carried out in the energy range of 1.3 to 5.0eV[85,86]. The electrolyte-electro-reflectance spectra for E‖b polarization here exhibited band-edge excitonic and inter-band transition features which were distinct from those for E⊥b polarization. Analysis of the polarization-dependent electrolyte-electro-reflectance spectra, and the structures of the excitonic and interband transitions of rhenium disulfide with optical polarization along the b-axis and perpendicular to the b-axis, clearly and accurately determined the energy positions (table 11). A mechanism involving field-lattice interaction was proposed in order to account for the in-plane anisotropy of the crystals.

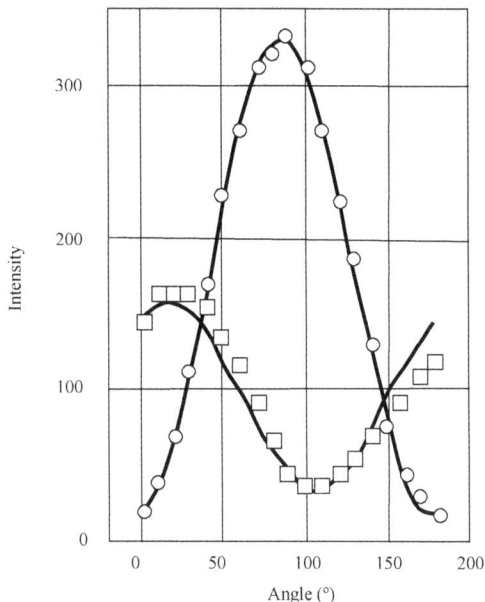

*Figure 26 Photoluminescence of ReS$_2$ at 110K Squares: $E_1^{ex}$, circles: $E_2^{ex}$*

The optical anisotropy of tungsten-doped layered crystals has been studied by means of polarization and temperature-dependent piezo-reflectance spectroscopy at 25 to 300K (figure 25). The direct band edge excitonic transitions, $E_1^{ex}$ for E‖b polarization and $E_2^{ex}$ for E⊥b polarization, were deduced from a detailed line-shape fit to the piezo-reflectance spectra[87]. The latter spectra revealed a slightly blue-shifted excitonic transition as a result of tungsten incorporation. The angular dependence of the excitonic feature amplitudes agreed with the Malus rule. This is the law, adumbrated by Malus in 1808 that, if natural light passes through two polarizing devices with their maximum transmission axes at an angle, θ, the exiting light intensity will be proportional to $\cos^2\theta$. The range of validity of the law has since been broadened as well as being applied to many photo-electric phenomena by directly relating the azimuthal angle to the non-equilibrium carrier concentration in an anisotropic semiconductor.

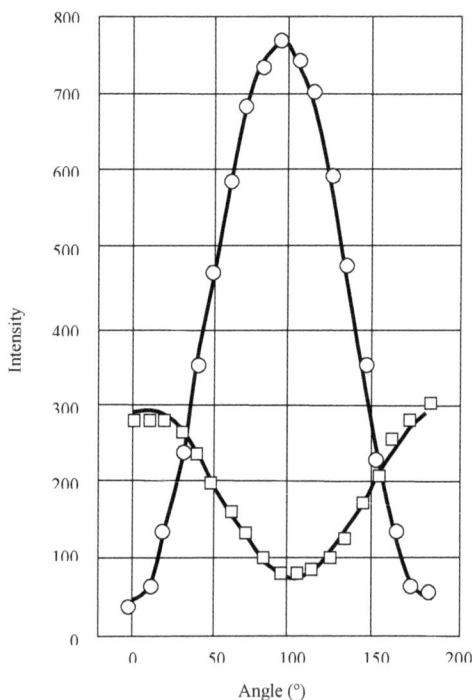

*Figure 27 Photoluminescence of gold-doped ReS$_2$ at 110K Squares: $E_1^{ex}$, circles: $E_2^{ex}$*

The excitonic transitions of the rhenium disulfide triclinic layered semiconductor were studied using polarized photo reflectance spectroscopy, with optical polarization along and perpendicular to the b-axis at 25 to 300K[88]. The low-temperature polarized photo reflectance spectra exhibited a prominent and enlarged excitonic series on the higher-energy side with respect to previously identified band-edge excitons, $E_1^{ex}$ and $E_2^{ex}$. Detailed line-shape analysis, together with the photosensitive characteristics of the excitonic sequence, furnished conclusive evidence that the band-edge excitons, $E_1^{ex}$ and $E_2^{ex}$, are interband excitonic transitions arising from differing causes. Proposed origins of the excitonic sequence ranged from non-bonding rhenium 5d $t_{2g}$ ($d_{xy}$, $d_{x2-y2}$) states to antibonding sulfur 3p $\sigma^*$ states. The sequence corresponded to the Rydberg series, beginning with the principal quantum number of n = 2.

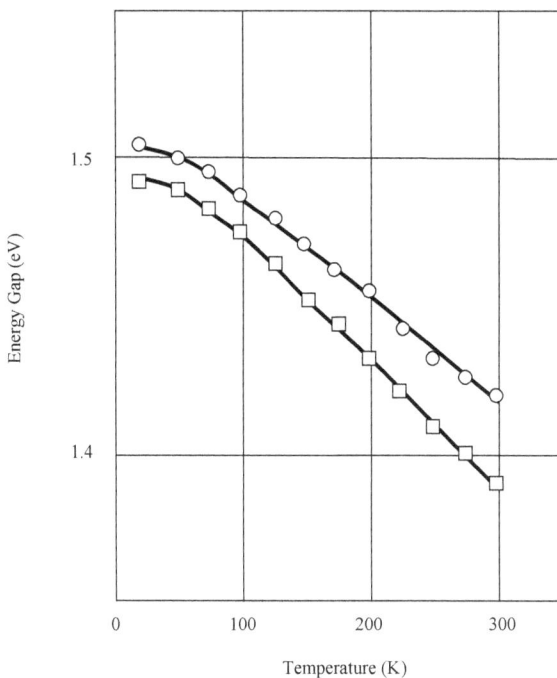

*Figure 28 Temperature dependence of the indirect energy gaps of gold-doped ReS$_2$ Circles: $E_{g\perp}$, squares: $E_{g\parallel}$*

Polarization-dependent thermo-reflectance and reflectance measurements were further made of the indirect band-gaps ($E_{g\parallel}$, $E_{g\perp}$), and direct band-edge excitonic transitions ($E_1^{ex}$, $E_2^{ex}$, $E_3^{ex}$, $E_4^{ex}$) at various polarization angles. Temperature-dependent polarization-dependent thermo-reflectance spectra were also determined between 55 and 300K. The amplitudes of the $E_1^{ex}$ and $E_2^{ex}$ transitions exhibited orthogonal characteristics and obeyed the Malus rule. A small blue-shift was observed in the indirect band-gaps and was attributed to doping effects[89].

The optical anisotropies of undoped and gold-doped rhenium disulphide have been investigated using polarization-dependent photoluminescence methods (figures 26 and 27)[90]. Because the excitonic transitions exhibit a strong polarization dependence in the near-infrared region, polarization-dependent measurements in the range from 0 to 180° were used to characterize the unique polarization properties of the materials and to identify the origin of the excitonic transitions. The variation in the amplitude of the photoluminescence excitonic transitions with various polarization characteristics obeys the Malus rule. When compared with undoped rhenium disulfide, the photoluminescence spectra of gold-doped material not only exhibit the main excitonic transitions near to the direct band-edge, but also extra transitions which are due to doping effects.

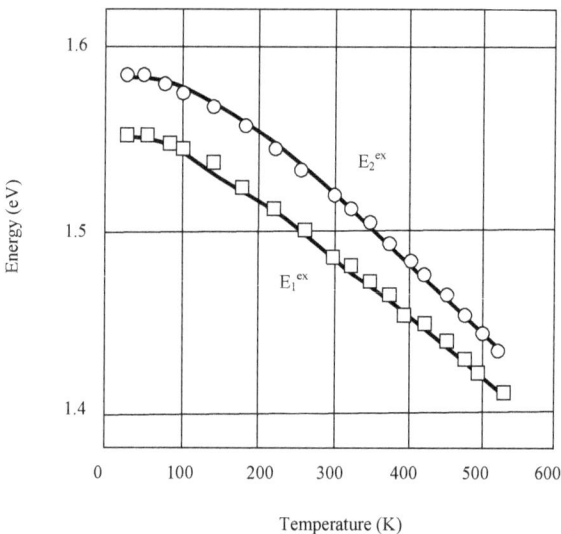

*Figure 29 Temperature dependence of the excitonic transition energies of ReS₂*

Gold-doped rhenium disulfide layer crystals (figure 28) have been grown using chemical vapour methods, with iodine as the transport agent. The thin samples were then investigated using polarization-dependent transmittance and photo-reflectance techniques at 20K[91]. The optical polarization was along and perpendicular to the crystal b-axis. The material exhibited a band-gap anisotropy with respect to the b-axis. Upon comparison with the undoped material, the indirect band-gap which was deduced from the polarization-dependent transmittance spectra exhibited a more pronounced red-shift than did the band-edge excitonic transitions deduced from the polarization-dependent photo-reflectance spectra. An optical study of the band-edge properties had previously been carried out using transmittance, photo-reflectance and piezo-reflectance methods. The polarized transmittance measurements indicated that the absorption edge of E∥b polarization exhibited a marked red-shift with respect to the E⊥b polarization[92]. The band-edge excitons, $E_1^{ex}$ and $E_2^{ex}$, were characterized by using polarized piezo-reflectance measurements. The polarization dependence of $E_1^{ex}$ and $E_2^{ex}$ provided conclusive evidence that the band-edge excitons are interband excitonic transitions arising from different causes. Higher-energy excitonic series in the layered compound were also studied by using low-temperature photo-reflectance measurements at 25K. Prominent and enlarged excitonic features located on the higher-energy side with respect to $E_1^{ex}$ and $E_2^{ex}$ were observed in the photo-reflectance spectra. The observed excitonic sequence again corresponded to the Rydberg series; beginning with the principal quantum number, n = 2. Room-temperature Hall-effect measurements of the chemical vapour iodine-grown gold-doped material revealed p-type semiconducting behaviour[93]. The electrical conductivities parallel to, and perpendicular to, the b-axis were investigated between 20 and 300K. In comparison with undoped material, gold-doped samples exhibited not only the main excitonic transitions, $E_1^{ex}$, $E_2^{ex}$, $E_3^{ex}$ and $E_4^{ex}$ near to the direct band-edge but also two extra transitions, $E_A^{ex}$ and $E_B^{ex}$, which were due to doping. The excitonic transitions exhibited a strong polarization dependence in the near-infrared region[94]. The variation in the amplitude of polarized piezo-reflectance excitonic transitions with various polarization characteristics again obeyed the Malus rule. In another polarization-dependent piezo-reflectance study[95], the band-edge exciton energies of rhenium disulfide as a function of temperature between 25 and 525K, were reported (figure 29, table 12).

*Table 12 Indirect band-gap and phonon energies obtained via polarization-dependent transmittance measurements*

| Material | Polarization | Energy Gap (eV) | Phonon Energy (meV) |
|----------|--------------|-----------------|---------------------|
| Undoped | E∥**b** | 1.35 | 25 |
| Undoped | E⊥**b** | 1.38 | 25 |
| Nb-doped | E∥**b** | 1.32 | 32 |
| Nb-doped | E⊥**b** | 1.35 | 27 |
| Mo-doped | E∥**b** | 1.31 | 32 |
| Mo-doped | E⊥**b** | 1.33 | 23 |

Other studies of the optical properties of gold-doped rhenium disulphide layer crystals, performed using polarization-dependent optical absorption and photoconductivity measurements, also showed that the absorption-edge shifted towards higher energies as the sample was thinned[96]. The room-temperature transition energies were 1.48eV for $E_1^{ex}$ and 1.516eV for $E_2^{ex}$. The $E_1^{ex}$ exciton predominated when the polarization was parallel to the b-axis of the layer crystal, while the $E_2^{ex}$ exciton was more prevalent when the polarization was perpendicular.

The optical properties of niobium-doped rhenium disulphide single crystals have been similarly studied by means of polarization-dependent transmittance, photoluminescence and piezo-reflectance measurements at 10 to 300K[97]. The indirect energy gap again exhibited a slight red-shift with respect to undoped samples (table 13). The low-temperature photoluminescence spectra contained two close direct band-edge excitonic peaks, together with two additional prominent features on the higher-energy side. The results agreed well with those of piezo-reflectance investigations of the same sample (table 14). When compared with the undoped sulfide, the excitonic transition energies remained essentially unchanged while the broadening parameter of the excitonic transition features increased slightly due to impurity-scattering.

*Table 13 Energy positions of the near band-edge excitonic transitions determined via polarization-dependent PzR measurements*

| Material | Temperature (K) | Component | Transition Energy (eV) |
|----------|-----------------|-----------|------------------------|
| Undoped | 30 | $E_1^{ex}$ | 1.554 |
| Undoped | 30 | $E_2^{ex}$ | 1.584 |
| Undoped | 300 | $E_1^{ex}$ | 1.484 |
| Undoped | 300 | $E_2^{ex}$ | 1.518 |
| Nb-doped | 20 | $E_1^{ex}$ | 1.555 |
| Nb-doped | 20 | $E_2^{ex}$ | 1.588 |
| Nb-doped | 300 | $E_1^{ex}$ | 1.485 |
| Nb-doped | 300 | $E_2^{ex}$ | 1.519 |
| Mo-doped | 30 | $E_1^{ex}$ | 1.554 |
| Mo-doped | 30 | $E_2^{ex}$ | 1.585 |
| Mo-doped | 300 | $E_1^{ex}$ | 1.485 |
| Mo-doped | 300 | $E_2^{ex}$ | 1.516 |

The optical Stark effect is a coherent light-matter interaction which describes the modification of quantum states by non-resonant light illumination in atoms, solids and nanostructures. Efforts have been made to exploit this effect in order to control exciton states and thus create ultra high-speed optical switches and modulators. These efforts have focused mainly on the optical Stark effect of only the lowest exciton state, due to a lack of energy selectivity, and results in the production of low degree-of-freedom devices. By applying a linearly polarized laser pulse to few-layer rhenium disulfide, in which the reduced symmetry leads to strong in-plane anisotropy of the excitons, control has been exercised over the optical Stark shift of two energetically separated exciton states[98]. In particular, the Stark effect of an individual state with varying light polarization can be selectively tuned. This is possible because each state exhibits an entirely distinct dependence upon light polarization, due to differing excitonic transition dipole moments. By using a varying pump-probe polarization configuration the excitons can be selectively shifted, based upon their anisotropic optical selection rules[99]. This offers a means for the energy-selective control of exciton states.

*Table 14 Broadening parameters of the near band-edge excitonic transitions determined via polarization-dependent PzR measurements*

| Material | Temperature (K) | Component | Broadening Parameter (meV) |
|----------|-----------------|-----------|----------------------------|
| Undoped | 30 | $\Gamma_1^{ex}$ | 6 |
| Undoped | 30 | $\Gamma_2^{ex}$ | 6 |
| Undoped | 300 | $\Gamma_1^{ex}$ | 31 |
| Undoped | 300 | $\Gamma_2^{ex}$ | 21 |
| Nb-doped | 20 | $\Gamma_1^{ex}$ | 7 |
| Nb-doped | 20 | $\Gamma_2^{ex}$ | 7 |
| Nb-doped | 300 | $\Gamma_1^{ex}$ | 32 |
| Nb-doped | 300 | $\Gamma_2^{ex}$ | 22 |
| Mo-doped | 30 | $\Gamma_1^{ex}$ | 8 |
| Mo-doped | 30 | $\Gamma_2^{ex}$ | 7 |
| Mo-doped | 300 | $\Gamma_1^{ex}$ | 35 |
| Mo-doped | 300 | $\Gamma_2^{ex}$ | 23 |

Because rhenium disulfide is a semiconducting layered transition metal dichalcogenide which has a stable distorted 1T phase, the low symmetry leads to an in-plane anisotropy of various properties. Thus a marked anisotropy is found in the Raman scattering response under linearly polarized excitation[100]. Polarized Raman scattering measurements permit the determination of the disulfide's crystallographic orientation by comparison with a direct structural analysis using scanning transmission electron microscopy. Analysis of the frequency difference between certain Raman modes provides a means for exactly determining layer thickness of up to four.

Surface-enhanced Raman scattering arising from two-dimensional layered materials provides a means for distinguishing chemical from electromagnetic mechanisms of enhancement. A chemical mechanism arises from charge interactions between the substrate and molecules, but it remains unclear as to how the electronic properties of the substrate are involved in charge interaction. Two-dimensional layered materials with anisotropic structures, such as the triclinic rhenium disulfide, may help to resolve the problem. Thus anisotropic Raman enhancement on few-layered rhenium disulfide is clarified by using copper phthalocyanine molecules as a Raman probe[101]. According to detailed Raman tensor analysis and density functional theory calculations, anisotropic charge interactions between the two-dimensional material and the molecules are responsible for the angular dependence of the Raman enhancement.

*Table 15 Raman active frequencies of ReS$_2$ under 633nm laser excitation*

| Symmetry | Sample | Frequency(/cm) | Origin |
|---|---|---|---|
| A$_g$ | bulk | 140.3 | out-of-plane vibration of Re atoms |
| A$_g$ | monolayer | 139.2 | out-of-plane vibration of Re atoms |
| E$_g$ | bulk | 151.3 | in-plane vibration of Re atoms |
| E$_g$ | monolayer | 153.6 | in-plane vibration of Re atoms |
| C$_p$ | bulk | 278.3 | in-plane and out-of-plane vibration of Re and S atoms |
| C$_p$ | monolayer | 278.3 | in-plane and out-of-plane vibration of Re and S atoms |
| E$_g$ | bulk | 307.8 | in-plane vibration of S atoms |
| E$_g$ | monolayer | 307.8 | in-plane vibration of S atoms |
| C$_p$ | bulk | 320.6 | in-plane and out-of-plane vibration of S atoms |
| C$_p$ | monolayer | 320.6 | in plane and out-of-plane vibration of S atoms |
| A$_g$ | bulk | 418.7 | out-of-plane vibration of S atoms |
| A$_g$ | monolayer | 419.3 | out-of-plane vibration of S atoms |

The Raman tensor of a crystal is the derivative of its polarizability tensor, and depends upon the symmetries of the crystal and the Raman-active vibrational mode. The intensity of a given mode is determined by the so-called Raman selection rule, which involves the Raman tensor and the polarization configuration. In anisotropic two-dimensional layered crystals, polarized Raman scattering is used to reveal crystal orientations but, due to its complicated Raman tensors and optical birefringence, the polarized Raman scattering of triclinic two-dimensional crystals has not been greatly studied. The anomalous polarized Raman scattering of two-dimensional layered triclinic rhenium disulfide reveals a large circular intensity differential of Raman scattering in samples of differing thickness[102].

The circular intensity differential, and the anomalous behavior of polarized Raman scattering, are attributed to the appearance of non-zero off-diagonal Raman tensor elements and the phase factor due to optical birefringence. This provides a technique for the identification of the vertical orientation of triclinic layered materials. The strong in-plane anisotropy of rhenium disulfide offers an additional physical parameter which can be tuned for advanced applications such as logic circuits, thin-film polarizers and polarization-sensitive photo-detectors. It is also advantageous for opto-electronic purposes, as rhenium disulfide is both a direct-gap semiconductor in the few-layer form (unlike $MoS_2$ or $WS_2$) and is stable in air (unlike black phosphorus). Raman spectroscopy is one of the most powerful characterization techniques for non-destructively probing the fundamental properties of two-dimensional material. A detailed study of the resonant Raman response of the eighteen first-order phonons in rhenium disulfide, for various layer thicknesses and crystal orientations, showed that rather than a general increase in the intensity of all of the Raman modes at excitonic transitions, each of the eighteen modes behaved differently relative to each other as a function of laser excitation, layer thickness and orientation[103]. This emphasizes the importance of electron-phonon coupling in this material.

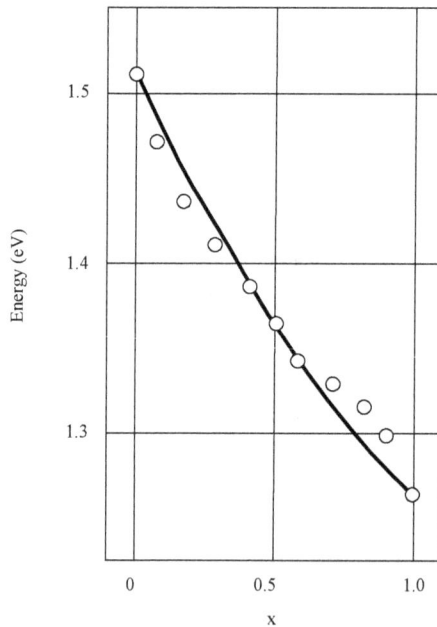

*Figure 30 Photoluminescence emission energy of bulk $ReS_{2(1-x)}Se_{2x}$*

Rhenium Disulfide          Materials Research Forum LLC
Materials Research Foundations **40** (2018)      doi: http://dx.doi.org/10.21741/9781945291920

This study also revealed a persistent and unrecognized error in the calculation of the optical interference enhancement of the Raman signal of transition-metal dichalcogenides on $SiO_2$/Si substrates. In the case of the present material, correcting this error was essential for the proper assessment of the resonant Raman behavior. A perturbation approach to calculating the frequency-dependent Raman intensities, based upon first-principles, demonstrated that in spite of neglecting excitonic effects, important trends in the Raman intensities of monolayer and bulk rhenium disulfide under various laser energies can be accurately described. Phonon dispersions, again calculated from first principles, could explain the possible origins of unexplained peaks which were observed in the Raman spectra. These included infrared-active modes, defects and second-order processes. As above, due to the lower structural symmetry and extraordinarily weak interlayer coupling of rhenium disulfide, it is possible to identify all eighteen of the first-order Raman active modes for bulk and monolayer samples[104]. With no van der Waals correction, local density approximation calculations could successfully reproduce all of the Raman modes. The calculations also suggested that there is no surface reconstruction effect, and an absence of low-frequency rigid-layer Raman modes below 100/cm. The combination of Raman spectroscopy with local density approximation calculations provides a general approach for predicting the vibrational and structural properties of two-dimensional layered materials of lower symmetry (table 15).

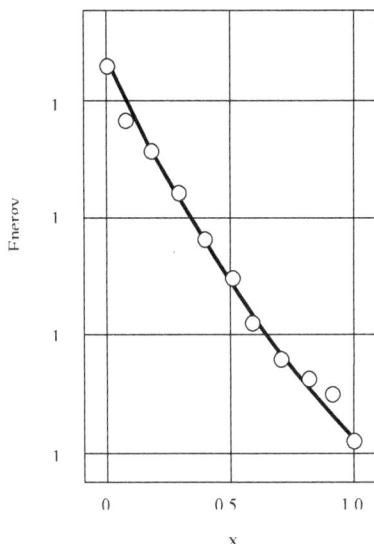

*Figure 31 Photoluminescence emission energy of monolayer ReS$_{2(1-x)}$Se$_{2x}$*

Unlike black phosphorus, which possesses poor environmental stability, rhenium disulfide has excellent stability but its electronic structure is only weakly dependent upon the number of layers; restricting the possibility of property-tuning and thus the number of device applications. One means of tuning properties, such as the optical band-gap, Raman anisotropy and electrical transport, is to alloy two-dimensional rhenium disulfide and selenide. The photoluminescence emission energy of $ReS_{2(1-x)}Se_{2x}$ monolayers, with x ranging from 0 to 1 in steps of 0.1, can be continuously tuned from 1.62 to 1.31eV (figures 30 to 34). The polarization behavior of the Raman modes, such as the $ReS_2$-like peak at 212/cm, shifts as the composition is changed[105]. The anisotropic electrical properties are maintained in the alloys, with a high electron mobility existing along the b-axis for all of the compositions.

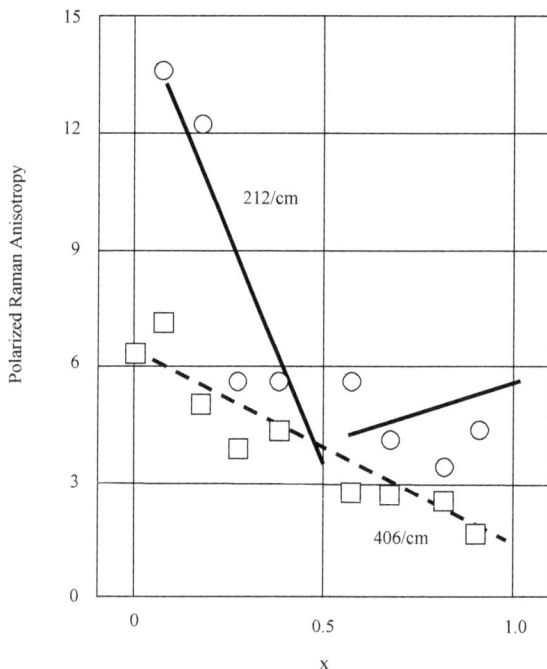

*Figure 32 Intensity anisotropy ratio for the 212 and 406/cm ReS$_2$ branches of ReS$_{2(1-x)}$Se$_{2x}$*

Materials Research Forum LLC
doi: http://dx.doi.org/10.21741/9781945291920

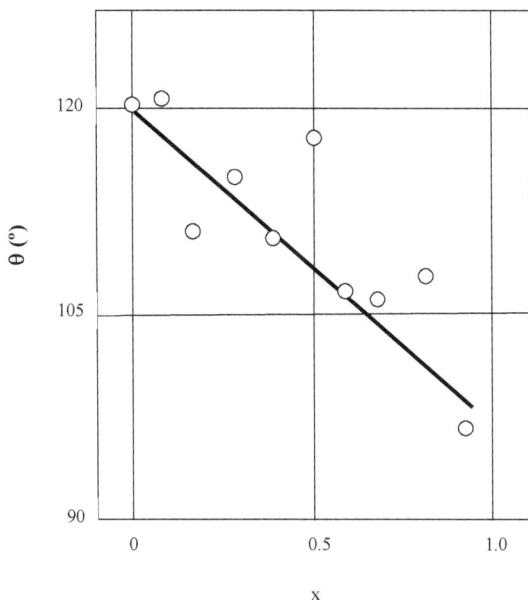

*Figure 33 Polarization angle for the 406/cm $ReS_2$ branch of $ReS_{2(1-x)}Se_{2x}$*

Two-dimensional materials in the series, $ReS_{2-x}Se_x$ ($0 \leq x \leq 2$), were prepared using chemical vapor transport and direct or indirect resonant emissions of the complete series of layers were simultaneously detected (figures 35 and 36) using polarized micro-photoluminescence spectroscopy when the sample thickness was greater than about 70nm.

When it was less, only three direct excitonic emissions ($E_1^{ex}$, $E_2^{ex}$, $E_S^{ex}$) were detected. In the case of thick (essentially bulk) samples, increased stacking of the monolayers flattened and shifted the valence-band maximum from the $\Gamma$-point to K- or M-related points[106]. This led to the coexistence of direct and indirect resonant light emissions from the c-plane.

The transmittance absorption edge of each bulk $ReX_2$ (a few microns thick) usually possessed a lower energy than those of direct $E_1^{ex}$ and $E_2^{ex}$ excitonic emissions involved in indirect absorption. This coexistence of direct and indirect emissions is a unique characteristic of two-dimensional layered semiconductors of triclinic low symmetry.

### Semiconduction

The valence-band maximum in pristine $ReS_2$ is made up of the 5d orbitals of rhenium atoms and the 3p orbitals of sulfur atoms. The conduction-band minimum is made up of the 5d orbitals of rhenium atoms. The material is an n-type direct band-gap semiconductor and the Fermi level is located at some 0.07eV below the bottom of the conduction band. An early *ab initio* band-structure calculation showed that rhenium disulfide is a semiconductor with an energy gap of about 1.0eV, while X-ray photo-emission spectra showed that it is a p-type semiconductor with an energy gap of about 1.5eV[107].

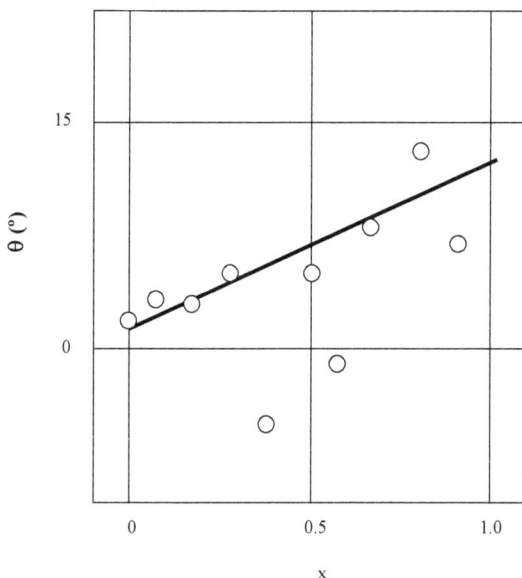

*Figure 34 Polarization angle for the 212/cm $ReS_2$ branch of $ReS_{2(1-x)}Se_{2x}$*

Semiconducting transition-metal dichalcogenides consist essentially of monolayers held together by weak forces via which the layers are vibrationally and electronically coupled.

The isolated monolayers exhibit changes in electronic structure and lattice vibration energy, generally including a transition from an indirect to direct band-gap. This variation is absent in rhenium disulphide, and the bulk here behaves as if electronically and vibrationally decoupled monolayers were stacked together[108]. So, from bulk to monolayer, rhenium disulfide maintains a direct band-gap and its Raman spectrum exhibits no dependence upon the number of layers. Further evidence of interlayer decoupling is provided by the insensitivity of the optical absorption and Raman spectrum to modulation of the interlayer distance by hydrostatic pressure (figure 37). This fundamental difference in behaviour is attributed to Peierls distortion of the 1T structure, which in turn prevents ordered stacking and minimizes interlayer overlap of the wave-functions. The absence of interlayer coupling offers the interesting advantage of being able to examine a two-dimensional system without having to produce monolayers.

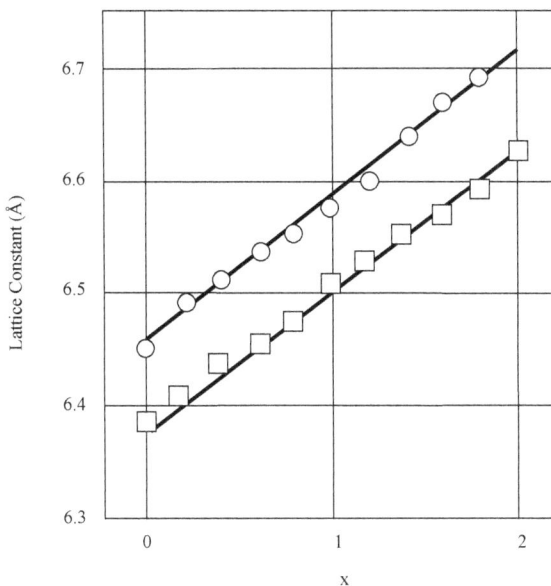

*Figure 35 In-plane lattice constants of the a- and b-axes of $ReS_{2-x}Se_x$ monolayers Circles: a-axis, squares: b-axis*

The results of a Raman spectra and first-principles theoretical study of lattice and electronic changes in compressed multilayer rhenium disulfide (figures 38 and 39) suggested the occurrence of an intralayer phase transition, followed by an interlayer transition. A transition from one indirect, to another indirect, band-gap at 2.7GPa was revealed by both high-pressure photoluminescence measurements and first-principles calculations[109]. The behavior was clarified by considering the fundamental relationship between lattice variations and electronic changes. By comparing the high-pressure behavior of $MoS_2$ and rhenium disulfide, it was demonstrated that interlayer coupling plays a critical role in determining the lattice and electronic structures in the compressed material.

In addition to previously-mentioned methods, two-dimensional rhenium disulfide nanosheets can also be produced by exfoliation using lithium intercalation. The electrochemical properties of such exfoliated disulfide nanosheets include low over-potentials of about 100mV and low Tafel slopes of 75mV/dec[110]. These features are attributed to the atomic structure of the superlattice 1T′ phase. The observation of band-gap photoluminescence indicates that these nanosheets retain their semiconducting properties, and disulfide nanosheets produced using this method offer unique photo-catalytic properties which are superior to those of other two-dimensional systems.

The absence of a direct-to-indirect band-gap transition in rhenium disulfide, upon going from the monolayer to the bulk material, causes it to be of particular interest among the semiconducting transition metal dichalcogenides. Functionalization of this material promises to offer a whole new generation of technological applications and devices. The structural, electronic and magnetic properties of rhenium disulfide monolayers, substitutionally doped at the sulfur or rhenium site have been predicted by means of first principles density functional calculations. It was concluded that substitutional doping depends greatly upon the growth conditions. Hydrogen, boron, carbon, selenium, tellurium, fluorine, bromine, chlorine, arsenic, phosphorus and nitrogen were considered as potential n-type and p-type dopants. Chlorine was concluded to be an ideal candidate for n-type doping while phosphorus appeared to be the best candidate for p-type doping of the sulfide monolayers. Predictions of the effect of doping with tungsten, chromium, cobalt, iron, manganese, nickel, copper, zinc, ruthenium, osmium, platinum, molybdenum, niobium, titanium or vanadium led to the conclusion that the latter four metals were easily incorporated into single-layer rhenium disulfide as substitutional impurities at the rhenium site under all of the growth conditions considered[111]. Energetic tuning of the chemical potentials of the dopant atoms makes it possible to dope rhenium disulfide with iron, cobalt, chromium, manganese, tungsten, ruthenium or osmium at the rhenium site. A clear trend is observed in the magnetic moments upon replacing a rhenium atom with a metallic atom. That is, depending upon the electronic configuration of the dopant atom, the net magnetic moment of the doped disulfide becomes either 0 or $1\mu B$. Among the metallic dopants, molybdenum was deduced to be the best candidate for the p-type doping of rhenium disulfide, due to its favorable energetics and interesting electronic properties.

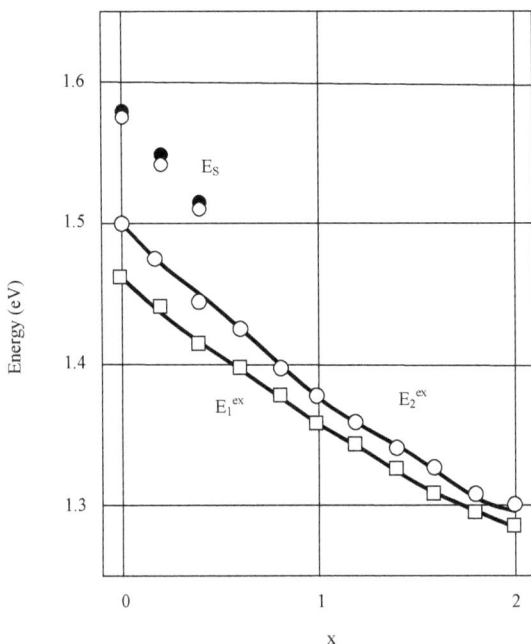

*Figure 36 Polarized μPL measurements of the emission energies of $ReS_{2-x}Se_x$ monolayers*

The layer-independent direct band-gap, 1.5eV, of rhenium disulfide implies a great future for the material in opto-electronic applications. The 1T′ structure, leading to anisotropic physical properties, also suggests that its associated electronic structure might harbour a non-trivial topology. An overall evaluation of the anisotropic Raman response and transport properties of few-layer rhenium disulfide field effect transistors shows that material which is exfoliated on $SiO_2$ behaves as an n-type semiconductor with an intrinsic carrier mobility greater than some $30cm^2/Vs$ at 300K. This increases to some $350cm^2/Vs$ at $2K^{112}$. Semiconducting behavior is observed at low electron densities, but, at high electron densities, the resistivity decreases by a factor of more than seven upon cooling to 2K, and exhibits a metallic $T^2$-dependence. This implies that the band structure of the 1T'-structured disulfide is susceptible to the effect of an electric field applied perpendicularly to the layers. An electric-field induced metallic state which is observed in transition metal dichalcogenides has been claimed to be the result of a percolation-type transition. Scaling analysis of the conductivity as a function of temperature and electron density shows that the metallic state of the disulfide results from a second-order metal-to-insulator transition, driven by electronic correlations. Such a gate-induced metallic state constitutes an alternative, to phase engineering, for producing ohmic contacts and metallic interconnects in devices based upon transition-metal dichalcogenides.

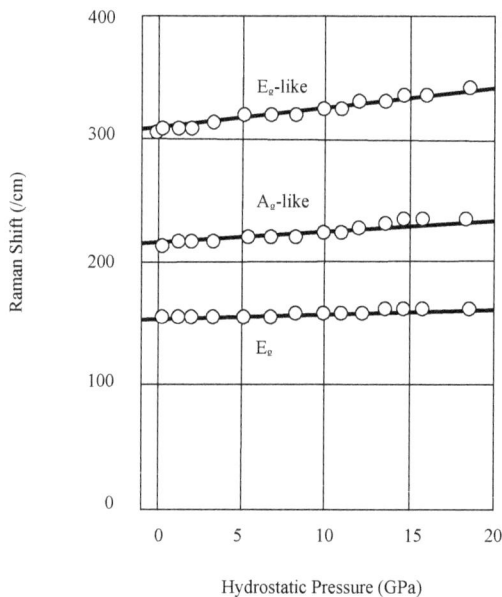

*Figure 37 Variation in the most prominent Raman peaks of bulk ReS$_2$ as a function of hydrostatic pressure*

It has been claimed that bulk crystals of rhenium disulfide are direct band-gap semiconductors and would be the ideal candidates, among all of the transition-metal dichalcogenides, for fabricating efficient opto-electronic devices. An indirect transition in the photoluminescence spectra has been reported however, and its energy is smaller than that of the supposed direct gap. The properties of ionic liquid gated field effect transistors have been exploited in order to investigate the gap structure of the bulk material. These transistors had an electron field effect transistor mobility of $1100cm^2/Vs$ at 4K, and hole transport at the surface of the material could be induced in order to make quantitative estimates of the smallest band-gap present in the material; regardless of its direct or indirect nature[113]. The value of the band-gap is 1.41eV; smaller than the 1.5eV direct optical transition but in good agreement with the energy of the indirect optical transition. This provided sufficient independent confirmation that the bulk material is an indirect band-gap semiconductor. Unlike other bulk semiconducting dichalcogenides such as $MoS_2$ and $WS_2$, rhenium disulfide field effect transistors fabricated on bulk crystals exhibit electroluminescence when driven in the ambipolar injection regime. This is probably because the difference between the direct and indirect gap is only 100meV.

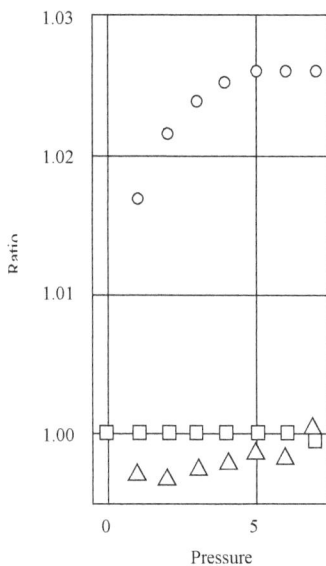

*Figure 38 Cell angle changes as a function of pressure Circles: $\alpha/\alpha_o$, squares: $\gamma/\gamma_o$, triangles: $\beta/\beta_o$*

Recalling that rhenium disulfide, like other group-VII transition-metal dichalcogenides, has a layered atomic structure with an in-plane motif of parallel diamond-shaped-chains, a combination of transmission electron microscopy and transport measurements revealed a direct correlation between electron-transport anisotropy in single-layered rhenium disulfide and the atomic orientation of the diamond-shaped chains. This was supported by density functional theory calculations[114]. A high chalcogen deficiency could induce structural transformation into a non-stoichiometric phase which was again strongly direction-dependent. This tunable in-plane transport behavior again suggests avenues of research for creating nano-electronic circuits in two-dimensional materials.

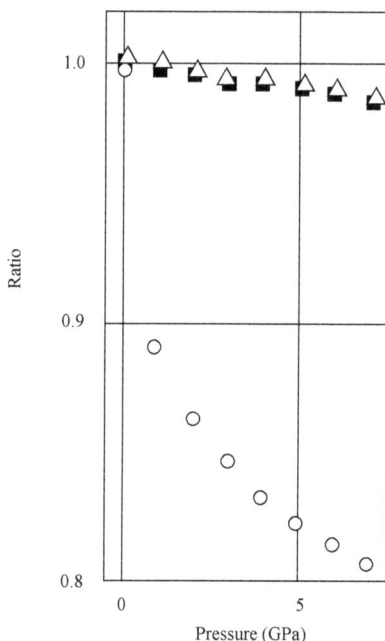

*Figure 39 Unit cell changes as a function of pressure Circles: $c/c_o$, squares: $a/a_o$, triangles: $b/b_o$*

Many-body perturbation theory calculations have been made of the excited-state properties of distorted 1T diamond-chain monolayer rhenium disulfide. The electronic self-energy appreciably enhanced the quasi-particle band-gap. The optical absorption spectra were governed by strongly bound excitons. Unlike hexagonal structures, the lowest-energy bright exciton of distorted 1T rhenium disulfide exhibited a perfect figure-of-eight shape polarization dependence[115]. These first-principles calculations were in excellent agreement with experimental data and were relevant to potential opto-electronic applications.

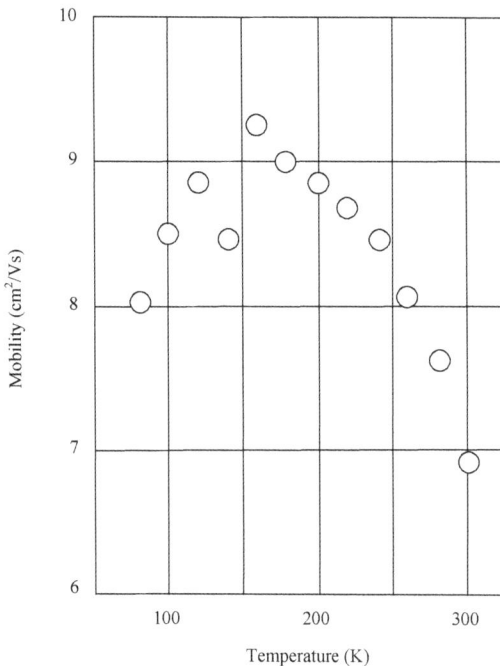

*Figure 40 Temperature dependence of the mobility of ReS₂*

The direct growth of high-quality rhenium disulfide atomic layers and nanoribbons has been demonstrated by using chemical vapor deposition. Chemical vapor deposited disulfide-based field-effect transistors exhibit n-type semiconduction, with a current on/off ratio of about $10^6$ and a charge-carrier mobility of some $9.3 cm^2/Vs$[116]. These results (figure 40) suggested that the quality of the chemical vapor deposited disulfide was comparable to mechanically exfoliated rhenium disulfide. This was confirmed by atomic force microscopic imaging, high-resolution transmission electron microscopic imaging and thickness-dependent Raman spectra data.

An electrical transport study of mono- and multi-layer rhenium disulfide with polymer electrolyte gating has shown that the conductivity of monolayers is completely suppressed at high carrier densities, thus making this disulfide the first known example of such a material[117]. Using dual-gated devices, it was possible to distinguish gate-induced doping from the electrostatic disorder which was introduced by the polymer electrolyte. Theoretical calculations and a transport model indicated that the observed conductivity suppression was due to the combination of a narrow conduction band and Anderson localization due to electrolyte-induced disorder.

A complete first-principles computational study of the effect of strain upon the anisotropic mechanical and electronic properties of rhenium disulfide monolayers showed that the anisotropic ratio of electron mobilities along the two principal axes is 2.36, while the ratio for hole mobility attains 7.76. A study of strain applied in various directions showed that the elastic modulus is greatest for the out-of-plane direction[118]. Straining in the a-direction introduced an indirect band-gap, while straining along the b- or c-directions did not have that effect. The carrier mobility could also be markedly improved by tensile straining in the c-direction. It can be concluded that rhenium disulfide monolayers have promising applications in nanoscale strain sensors and conductance-switch field effect transistors.

The structural, electronic and optical properties of monolayer rhenium disulfide have been investigated using quantum mechanical calculations. The calculated electronic band-gap was 1.43eV; with a non-magnetic ground state[119]. The calculated dopant substitutional energies under rhenium-rich and sulfur-rich conditions show that it is possible to synthesize transition metal doped monolayer systems. The presence of dopant ions such as vanadium, chromium, manganese, iron, cobalt, niobium, molybdenum, tantalum or tungsten in rhenium disulfide monolayers greatly modifies the electronic ground states; consequently introducing defect levels and modifying the density-of-states profile. Manganese-doped structures suffer a very minute reduction in the electronic band-gap. A ferromagnetic or a non-magnetic ground state configuration was obtained, depending upon the choice of dopant. Doping with chromium, iron or cobalt results in a ferromagnetic ground state configuration of the disulfide structure. The calculated absorption and reflectivity spectra show that this class of dopant causes a general increase in the absorption spectral peaks, but has only a minute effect upon the reflectivity. Optical anisotropy is observed, depending upon whether the polarization direction in the xy-plane is parallel or perpendicular.

*Table 16 Lattice parameters of $ReS_{2-x}Se_x$ layered compounds*

| Composition | a(Å) | b(Å) | c(Å) | α(°) | β(°) | γ(°) | V(Å³) |
|---|---|---|---|---|---|---|---|
| $ReS_2$ | 6.450 | 6.390 | 6.403 | 105.49 | 91.32 | 119.03 | 217.97 |
| $ReS_{1.8}Se_{0.2}$ | 6.496 | 6.412 | 6.449 | 104.71 | 91.93 | 119.12 | 223.01 |
| $ReS_{1.6}Se_{0.4}$ | 6.517 | 6.434 | 6.488 | 104.77 | 91.77 | 119.19 | 225.73 |
| $ReS_{1.4}Se_{0.6}$ | 6.539 | 6.460 | 6.517 | 104.85 | 91.65 | 119.19 | 228.41 |
| $ReS_{1.2}Se_{0.8}$ | 6.556 | 6.485 | 6.538 | 104.71 | 91.68 | 119.22 | 230.74 |
| ReSSe | 6.575 | 6.508 | 6.581 | 104.70 | 91.54 | 119.20 | 233.95 |
| $ReS_{0.8}Se_{1.2}$ | 6.602 | 6.527 | 6.613 | 104.65 | 91.61 | 119.11 | 236.99 |
| $ReS_{0.6}Se_{1.4}$ | 6.640 | 6.546 | 6.651 | 104.59 | 91.78 | 119.14 | 240.28 |
| $ReS_{0.4}Se_{1.6}$ | 6.666 | 6.566 | 6.676 | 104.58 | 92.10 | 119.00 | 242.95 |
| $ReS_{0.2}Se_{1.8}$ | 6.690 | 6.591 | 6.703 | 104.60 | 92.18 | 118.85 | 246.03 |
| $ReSe_2$ | 6.713 | 6.623 | 6.740 | 104.59 | 92.28 | 118.79 | 249.53 |

Single crystals of molybdenum-doped rhenium disulphide have been grown by chemical vapour transport, using bromine as the transport medium[120]. Monocrystalline platelets, 100μm in thickness and with a surface area of up to 5mm x 5mm, were obtained. X-ray diffraction patterns showed that the doped crystals crystallized with a triclinic layered structure. Hall-coefficient measurements indicated that the samples were n-type in nature. The effects of doping were characterized by performing temperature-dependent conductivity, optical absorption and piezo-reflectance measurements. The activation energies of impurity carriers increased with dopant addition, and the indirect energy-gap of doped samples exhibited a slight red-shift. The direct band-edge excitonic transition energies meanwhile remained unchanged while the broadening parameter of the excitonic transition features increased due to impurity scattering.

The optical properties of rhenium disulfide were characterized by performing polarized thermo-reflectance measurements, between 25 and 300K, on single crystals which had been grown by chemical vapor transport with $Br_2$ as the transport medium. The as-grown disulfide exhibited two different structural phases following crystallization: normal triclinic layer and tetragonal. The polarized thermo-reflectance measurements were performed along, and perpendicular, to the b-axis for both layered and tetragonal crystals. The occurrence of structural changes was attributed to atomic bonding deformation along the b-axis; this being parallel to the $Re_4$ parallelogram diamond chains[121].

Single crystals of $ReS_{2-x}Se_x$ solid solutions were also grown by chemical vapor transport, using $Br_2$ as the transport medium[122]. Analysis of X-ray patterns showed that the crystals were single-phase and had a triclinic layered structure, the lattice parameters of which were determined (table 16). The optical absorption edge was measured on basal planes at room temperature and the results revealed that the materials were indirect semiconductors, the energy gaps of which were determined (table 17). The band-gap energy varied smoothly with the amount of selenium addition, indicating that the nature of the band-edges is similar for the end-members and for the intermediate compositions.

*Table 17 Compositional dependence of indirect gaps and energies of monocrystalline $ReS_{2-x}Se_x\, E_x = E_o + bx + cx^2$*

| Parameter | Temperature (K) | $E_o(eV)$ | b | c |
|-----------|----------------|-----------|-------|-------|
| $E_g$ | 300 | 1.367 | -0.125 | 0.021 |
| $E_1^{ex}$ | 25 | 1.558 | -0.13 | 0.021 |
| $E_2^{ex}$ | 25 | 1.59 | -0.135 | 0.021 |
| $E_g^{ex}$ | 300 | 1.52 | -0.15 | 0.021 |

A systematic scanning tunnelling microscopic study was made of the local photovoltaic properties of rhenium disulfide. The tunnelling junction of the scanning tunnelling microscope was optically illuminated during the tunnelling process. The phase-sensitive photo-induced tunnelling current was then monitored as a function of the wavelength and surface topography. In order to improve the performance of disulfide solar cells, the samples were treated with $NaI/I_2$ and ethylenediaminetetra-acetic acid solutions. In comparison to untreated sample, ethylenediaminetetra-acetic acid-treated samples exhibited a factor of 8 to 10 increase in the photo-induced tunnelling current over the entire spectral range. The $NaI/I_2$-treated samples exhibited a factor of 2 to 3 increase[123].

During the electrochemical oxidation of rhenium disulfide crystals, the anodic photocurrent in aqueous electrolytes can increase by up to 140 times, accompanied by the formation of an unstable $Re_2O_7$ surface layer and a change in the near-ultraviolet photocurrent spectrum[124]. This is attributed to the formation of a heterojunction between the rhenium disulfide and the $Re_2O_7$. Upon adding a $I^-/I_3^-$ redox couple, a transient electrochemical solar cell is obtained. Further study of the interfacial quantum processes has shown that the enhancement is due to light-collection by scattering processes in the highly refractive $Re_2O_7$ film and at the epoxy insulation layer.

Rhenium disulfide, being a two-dimensional characterized by having low symmetry and containing one-dimensional metallic chains within its planes - plus extremely weak interlayer bonding - can be expected to exhibit unusual transport properties. Using the time-domain thermo-reflectance method, the room-temperature thermal conductivity of monocrystalline rhenium disulfide, with its distorted 1T structure, has been determined[125] for in-plane directions which were parallel or perpendicular to the Re-chains, and for the through-plane direction. The disulfide was prepared in the form of flakes with a thickness of 60 to 450nm by means of micromechanical exfoliation. The crystal orientations were determined by means of polarized Raman spectroscopy. The in-plane thermal conductivity of 70W/mK was higher along the Re-chains, as compared to that (50W/mK) transverse to the chains. Due to the weak interlayer bonding, the through-plane thermal conductivity of 0.55W/mK was perhaps the lowest found for a two-dimensional material. When compared to the in-plane directions this amounted to anisotropies of 130 and 90.

It can be seen that one important factor in the promised usefulness of rhenium disulfide is the assumption that its direct band-gap is independent of the thickness of the material. This is the consensus opinion, but there are voices of dissent. An early report [126] pointed out that bulk $MoS_2$, the typical layered transition-metal dichalcogenide, is an indirect band-gap semiconductor but upon reducing its thickness from a slab form to a monolayer, it undergoes a transition to a direct band-gap semiconductor. First-principles calculations showed that quantum confinement in layered d-electron dichalcogenides resulted in a tuning of the electronic structure. Application to related ($WS_2$, $NbS_2$, $ReS_2$) nanolayers showed that the first of these exhibits similar electronic properties while the others remained metallic, regardless of the slab thickness. The nature, direct or indirect, of the band gap of bulk, few-layer and monolayer forms of rhenium disulfide is related to its relatively weak interplanar interaction. The question of whether a transition from an indirect to a direct band gap occurs with decreasing thickness is somewhat unsettled. Direct determination of the valence-band structure of bulk material, using high-resolution angle-resolved photo-emission spectroscopy, revealed[127] a clear in-plane anisotropy that is due to the existence of chains of rhenium atoms having a strongly directional effective mass which is greater ($2.2m_e$) in the direction orthogonal to the rhenium chains than along them ($1.6m_e$). Interplanar interaction results in a difference of 100 to 200meV between the valence-band maxima at the Z-point (0,0,12) and $\Gamma$-point (0,0,0) of the three-dimensional Brillouin zone. This results in a direct gap at Z and a nearby but larger gap at

$\Gamma$; implying that the bulk-material gap is marginally indirect in nature and thus difficult to characterise. An angle-resolved photo-emission spectroscopic investigation[128] of the band structure of rhenium disulfide revealed a large number of narrow valence bands which were attributed to the combined effect of structural distortions and spin-orbit coupling. This led to marked in-plane anisotropy of the electronic structure, with quasi one-dimensional bands indicating predominant hopping along zig-zag rhenium-atom chains. This did not persist up to the top of the valence band. A more three-dimensional behaviour occurred there, with the fundamental band-gap being located away from the Brillouin-zone center. The experimental data were in good agreement with density-functional theory calculations, thus helping to clarify the bulk electronic structure of the disulfide and its change during thinning to a single layer. As a further check, the disulfide surface was electron-doped by depositing low concentrations of rubidium atoms which populated the conduction-band states. The angle-resolved photo-emission spectra from as-cleaved and rubidium-doped samples indicated that the new states appeared at about 1.2eV above the valence-band maximum, located at the $\bar{\Gamma}$-point of the surface Brillouin zone, and were considered to be occupied conduction-band states. Photon-energy dependent angle-resolved photo-emission data indicated that the conduction-band minimum as well as valence-band maximum exhibited appreciable dispersion along $k_z$. The conduction and valence bands dispersed in opposite senses, meaning that a direct band-gap with a high joint density of states persisted in the bulk ReS$_2$, if only located at the Brillouin zone boundary along $k_z$, rather than at the Brillouin zone center. The measurements could not entirely exclude the possible existence of a slightly smaller indirect band gap which was located away from high-symmetry lines. The nature of the optical band gap is also in dispute, with the existence of an indirect band gap being suggested by optical measurements[129]. The band gap of about 1.2eV which was deduced from the angle-resolved photo-emission measurements was some 300meV smaller than the measured optical band gap. The present surface-doping was expected to create a near-surface downward band-bending rather than causing rigid band-filling of the conduction-band states. This was however to be expected to lead to an increased band gap due to quantum-confinement of the conduction-band states and was inconsistent with the observed three-dimensional dispersions of the conduction-band states. The smaller band gap which was observed could reflect an increased electronic screening that was due to the high near-surface electron density. This would create a marked renormalization of the electronic band gap with respect to its value in the undoped semiconductor and could indicate the presence of quite strongly-bound excitons, even in the bulk diselenide.

First-principles density functional theory calculations of the ohmic nature of the top-contact formed by monolayer rhenium disulfide with gold, silver, platinum, nickel, titanium or scandium clarified the potential barrier, charge transfer and atomic orbital overlap characteristics at the disulfide/metal interface in terms of van der Waals forces[130]. The results explained how efficiently carriers can be injected from the metal contact and into the monolayer disulfide channel. The disulfide was physisorbed on gold and silver, leading to minor perturbation of its electronic structure and leading to a greater Schottky contact and a tunnel barrier at the interface. On the other hand, the disulfide was chemisorbed on titanium and scandium and the bonding markedly disturbed the electronic structures. The behaviour was purely ohmic. Finally, the bonding of rhenium disulfide on platinum and nickel fell between the above extremes and led to an intermediate behavior. A similar study[131] again showed that the monolayer disulfide exhibited a lower bonding with gold and silver, leading to just slight perturbation of its electronic structures and forming a larger Schottky contact and higher tunnel barrier at the interface, as compared with the bonding of the disulfide on the platinum surface.

Rhenium disulfide is a promising two-dimensional material for opto-electronic devices because of an excellent photonic response which arises from the insensitivity of its band-gap energy to the layer thickness. A theoretical calculation[132] of the electrical band structures of monolayer, bilayer and trilayer disulfide, plus experiment, indicated a work function of 4.8eV. This value was independent of the layer thickness. Evaluation of the contact resistance of a rhenium disulfide field-effect transistor, using a Y-function method and various metal electrodes including graphene, showed that the disulfide channel was a marked n-type semiconductor. A lower work function than that of metals thus tended to lead to a lower contact resistance. Graphene electrodes which were not chemically or physically bonded to the disulfide exhibited the lowest contact resistance, regardless of the magnitude of the work function. This suggested that there was an appreciable Fermi-level pinning effect acting at the disulfide/metal interface. An asymmetrical Schottky diode device was also prepared by using titanium or graphene as ohmic contacts, and platinum or palladium as Schottky contacts.

*Magnetic*

The electronic and magnetic properties of transition-metal atoms (cobalt, copper, iron, manganese, nickel), adsorbed on a rhenium disulfide monolayer, have been investigated using first-principles calculations. Magnetism appears in the case of cobalt, iron and manganese. The manganese-adsorbed system has the most stable structure. Further study of the two-Mn-adsorbed system showed that the interaction between two manganese atoms is always ferromagnetic, but the interaction is suppressed by increasing the Mn-Mn distance. This could be well-described by using a simple Heisenberg model which was based upon the Zener-Ruderman-Kittel-Kasuya-Yosida theory[133]. The cobalt-doped system had the largest magnetic moment. Further study of interactions in a two-Co-doped system showed that, as in the case of manganese, the interaction between two cobalt atoms was always ferromagnetic, but the interaction weakened with increasing Co–Co distance. This again was well-described by a simple Heisenberg model based upon the Zener theory[134]. The magnetic properties of rhenium disulfide monolayers, doped with non-metallic elements, have also been studied using first-principles methods. Various dopants (boron, carbon, chlorine, fluorine, nitrogen, oxygen) and doping sites were considered. As in the case of transition-metal atoms, a magnetic behavior appears in the boron-, carbon-, fluorine- and oxygen-doped systems[135]. The calculated binding energies show however that the boron-doped system constitutes the most stable system among the four magnetic materials. Ferromagnetic interactions were studied in two boron-doped rhenium disulfide monolayers. Upon increasing the B-B distance, both ferromagnetic and non-magnetic behaviours were found. It is proposed that the ferromagnetic coupling originates from a p-d exchange-like p–p coupling interaction. Finally, the magnetic properties of rhenium disulfide monolayers doped with non-magnetic metals (silver, aluminium, lithium, magnesium, sodium) have been studied using a first-principles method. Magnetic behavior appeared in the aluminium- and magnesium-doped systems. The calculated binding energies showed that the aluminium-doped system was more stable than the magnesium-doped one[136]. Study of the ferromagnetic interaction in two Al-doped rhenium disulfide monolayers showed that, as the Al–Al distance increased, both ferromagnetic and non-magnetic states were found. The ferromagnetic coupling was suggested to arise from a p-d exchange-like p-p interaction.

In detailed studies, a few-layer (8–9) $ReS_2$ flake was exfoliated onto a sapphire substrate from a bulk monocrystal a using polydimethylsiloxane stamp-assisted mechanical technique. The energies of the anisotropic excitons were determined using static spectroscopy. The absorption resonances of the two lowest excitons (1.531 and 1.566eV) exhibited characteristic dependences on the light polarization; the polarization angle with respect to the b-axis. Pump-probe experiments were performed by using linearly polarized pulses that were centered around 800nm. The pulse spectrum included the above excitons, so as to stimulate them simultaneously. A weak portion of the split beam served as the probe pulse, and the detection wavelength was set to the center (800nm) of the excitons in order to maximize the beat amplitude. The differential transmittance was measured as a function of the pump-probe time delay. The differential transmittance signal was defined as $\Delta T/T$, where T was the probe intensity without a pump and $\Delta T$ was the pump-induced change in the probe intensity. The pump fluence was fixed at $15\mu J/cm^2$ and the sample temperature was 79K. The differential transmittance exhibited periodically oscillating signals, after 0.15ps, which could be described by a damped harmonic oscillator function. The oscillation period of 116fs agreed exactly with the energy-splitting between the above excitons. Such an oscillation behavior was typical of the quantum coherence that arose from exciton quantum-beats. No quantum beats were observed when the pump resonantly excites one of the two excitons, thus indicating that the coherent oscillations do not originate from a single excitonic feature. Unlike energetically degenerate valley excitons, the de-phasing dynamics of the present 2 excitons should include fast inter-excitonic relaxation from the higher to the lower exciton absorption resonance. The observations implied that the excitonic quantum coherence in $ReS_2$ imparted a marked independence with respect to relaxation dynamics and temperature variations. It is known that quantum beats can be generated only in a coupled three-level system with a common state. On the other hand, independent two-level oscillators can produce a quantum beat-like signal, known as polarization interference, which arises from electromagnetic interference at the detector. Because polarization interference has the same temporal shape as quantum beats, ingenious methods have been devised for distinguishing between them. Some are based upon time-resolved or spectrally-resolved analyses of 3-pulse 4-wave mixing experiments. This was not true of the 2-pulse pump-probe technique, and so a theoretical analysis was here based upon the semi-classical Liouville-von-Neumann equation. The key result was that the two-pulse pump-probe technique does not permit polarization interference and so no additional steps are required in order to confirm the occurrence of genuine quantum beats, unlike the case of 3-pulse 4-wave-mixing experiments. The 2-pulse pump-probe technique is therefore dependable and convenient for identifying quantum beats. The present theoretical analysis confirmed that the observed quantum beats for the studied excitons originated from quantum coherence in a coupled 3-level configuration. The beating is clearly visible only at polarization angles of 90 to 150°, and its clear angle-dependence suggests that the quantum beats can be switched on and off simply by altering the laser optical polarization. A deeper understanding of the angle-dependence of the quantum beats can be obtained by using a simple theoretical model which assumes that the beat amplitude is proportional to the product of the exciton spectral weight. This amplitude exhibits peaks at 50° and 140°, and indicates that the quantum beats have a relatively large intensity when the population of the two excitons is balanced. The magnitude of the two peaks is different because the polarization directions of the excitons (that direction at which the spectral weight attains its maximum) are not exactly mutually orthogonal. When compared with experiment, the measured peak amplitude in one region agreed well with the theory while too small a value was found in another region. The latter failure was attributed to the possibility that the most prominent phonon mode was strongly anisotropic and that the polarization angle at which the phonon response was maximum coincided with the region of poor prediction. Weak beats in that could then arise from an anisotropic exciton–phonon scattering which led to extremely fast de-phasing. Another possible explanation was that the angle between the excitons in the poor-prediction region was smaller than that in the other region. Given that the anisotropic transitions of the excitons have similar polarization dependences in momentum space, the electron and hole states of one exciton can be located quite close to those of the other exciton in the region of theory failure. This could lead to fast de-phasing via rapid relaxation from one exciton to the other. Overall, the observations were somewhat surprising in that the anisotropic excitons exhibited entirely different optical selection rules and orientations. In spite of the presence of inter-excitonic relaxation and a relatively high temperature, the measured de-phasing time of 100 to 200fs was comparable to that of the valley excitons in $MX_2$ compounds at 10K, thus implying a marked stability of exciton coherence in $ReS_2$. Besides the considered excitons were close to 800nm in the spectral domain, it was moreover possible to generate exciton quantum beats simply by using the output of commercial lasers, without any wavelength conversion.

Quantum beats, periodic oscillations arising as they do from coherent superposition states, permit the investigation of novel coherent phenomena. Because of their strong Coulomb interactions and reduced dielectric screening, the two-dimensional transition metal dichalcogenides exhibit strongly bound excitons in single structures or heterostructures. Quantum coherence between excitons is however not a common effect. Exciton quantum beats in atomically thin $ReS_2$ are observed[137] to modulate further the intensity of the quantum-beat signal. Linearly polarized excitons behave like a coherently coupled three-level system and exhibit quantum beats even though they exhibit anisotropic exciton orientations and obey optical selection rules. Theoretical work shows that the observed quantum beats arise from pure quantum coherence phenomena rather than from classical interference effects. The ON/OFF quantum beats can also be modulated, merely by means of laser polarization.

## Applications

It is seen from the above that rhenium disulfide, as a two-dimensional group-VII transition-metal dichalcogenide, possesses structural and vibrational anisotropy, layer-independent electrical and optical properties and exhibits metal-free magnetism. These properties are unusual when compared with more widely-used group-VI transition-metal dichalcogenides such as $MoS_2$, $MoSe_2$, $WS_2$ and $WSe_2$. The present disulfide is expected to be especially useful when combined with isotropic transition-metal dichalcogenides from group VI[138].

The present disulfide unit cell has a basic hexagonal symmetry in which rhenium atoms are grouped into parallelograms of 4 rhenium atoms which allow great latitude in the introduction of planar anisotropies into composite heterostructures. Other two-dimensional materials tend to exhibit properties which are dissimilar to those of the bulk form. Both two-dimensional and three-dimensional forms of $ReS_2$ exhibit however the same physicochemical properties. The optical and opto-electronic properties of two-dimensional layered semiconductors depend upon the number of layers, but $ReS_2$ possesses almost layer-independent opto-electronic properties. Due to dimensional-confinement effects, and to modulation of the band structure, there is an essentially unavoidable change in optical absorption – in going from the infra-red to the ultra-violet - in other two-dimensional materials. The present disulfide is also mechanically flexible, and interacts strongly with incident light; potentially leading to increased photon absorption and electron–hole pair generation. Due to the van der Waals interactions between layers, and the absence of surface dangling bonds, it is possible to combine it with other materials without the usual limitations imposed by lattice-matching. In terms of device fabrication, $ReS_2$ could potentially replace graphene because the latter's lack of a band-gap makes it rather useless when a semiconductor is required for an application. Rhenium disulfide also promises to be the basis of more energy-efficient molecular-scale digital processors than is silicon; in opto-electronic device fabrication, it avoids the limitations of bulk silicon due to its high absorption efficiency at broadband wavelengths.

In spite of the extensive work expended on obtaining high-quality rhenium disulfide nanoflakes, uniform in-plane growth remains difficult because of the unique decoupling between layers. Mechanical exfoliation is a typical method used for obtaining high-quality nanoflakes, but the resultant monolayer or few-layer disulfide product is always very small and little suitable for nanoscale devices. Chemical vapor deposition is an efficient means for achieving large-area monolayer material but the synthesized monolayers tend to contain high levels of vacancies due to insufficient reaction at the relatively low temperatures involved. Highly crystallized pyramid-like few-layer flakes have been prepared by using rhenium powder as a source, but the growth rate was limited by the high melting point of the metal. Strong interlayer decoupling of $ReS_2$ also made out-of-plane growth predominate. Substrates having a low surface energy can facilitate atomic migration along in-plane directions and thus help to obtain uniform films with very flat surfaces. Epitaxial growth of continuous monolayer rhenium disulfide films on mica substrates has been successfully achieved[139] by means of chemical vapour deposition. Continuous multilayer disulfide films could be produced by extending the growth period. The growth of the disulfide films was explained in terms of Stranski-Krastanov theory. Field effect transistors which were based upon multilayer rhenium disulfide films exhibited a typical n-type semiconducting behaviour with a carrier density of $0.27cm^2/Vs$ and an ON/OFF ratio of about 4 x $10^3$. The photo-response of a phototransistor could attain up to 0.98A/W with a light intensity of $0.56mW/cm^2$.

### Electronic Devices

There is huge current interest in the novel device applications of low-symmetry two-dimensional materials. These include black phosphorus and its arsenic alloys, compounds having a black-phosphorus like structure, such as the monochalcogenides of group-IV elements such as germanium and tin, as well as the class of low-symmetry transition metal dichalcogenide materials such as rhenium disulfide and rhenium diselenide. The two-dimensional transition metal dichalcogenides offer a number of attractive features for exploitation in the next generation of electronic and opto-electronic devices, such as field effect transistors, photovoltaic cells, light-emitting diodes, photo-detectors, lasers and integrated circuits.

Exfoliated few-layer dual-gated rhenium disulfide field effect transistors have been fabricated. These exhibited n-type behavior, with a room-temperature on/off ratio of $10^5$. Many of the devices had a maximum intrinsic mobility of $12cm^2/Vs$ at room temperature and $26cm^2/Vs$ at 77K. The Cr/Au-ReS$_2$ contact resistance, as determined using the transfer length method, was gate-bias dependent and ranged from 175 to 5k$\omega\mu$m[140]. An exponentially measured dependence upon the back-gate voltage indicated the presence of Schottky barriers at the source and drain contacts. Dual-gated ReS$_2$ field effect transistors exhibited current saturation, voltage gain and a sub-threshold swing of 148mV/decade.

The fabrication of field effect transistors has also been achieved by encapsulating rhenium disulfide nanosheets within a high-κ Al$_2$O$_3$ dielectric. Few-layer ReS$_2$ nanosheets were deposited on SiO$_2$/Si substrates in a chemical vapor deposition furnace by using ReO$_3$ and sulfur as the source materials. The disulfide had crystallized in the familiar distorted 1T structure, with clusters of Re$_4$ units forming a one-dimensional chain in each monolayer. Photoluminescence spectra for 3 layers and for the bulk revealed the expected negligible band-gap change from 1.54 to 1.50eV. The presence of differing thicknesses could be detected by optical contrast studies, and precisely determined using atomic force microscopy. Again as expected, Raman spectroscopy revealed no thickness-dependence due to the decoupling of lattice vibrations between adjacent layers. Low-temperature transport measurements detected a direct metal-to-insulator transition which arises from strong electron-electron interactions[141].

A unique feature of two-dimensional materials is their ability to form heterojunctions which exhibit good interface quality without any lattice mismatch problem. A

ReS$_2$/ReSe$_2$ van der Waals heterostructure was shown[142] to possess gate-tunable diode properties, with a maximum rectification ratio of 3150. Under illumination, it exhibited a photovoltaic effect with an efficiency of about 0.5%.

Top-gate field-effect transistors were fabricated using standard e-beam lithography and metal deposition. A 30nm-thick Al$_2$O$_3$ layer was deposited via atomic layer deposition, and the oxide served as the top-gate dielectric. The Fermi level of the encapsulated ReS$_2$ channel could be tuned by changing the top-gate or back-gate voltage applied to the degenerately-doped silicon substrate. In order to study the gate modulation of the ReS$_2$ nanosheets, source-drain current–voltage curves were determined. The source-drain current varies linearly with the source-drain voltage, thus indicating well-developed contact between the electrodes and the ReS$_2$ channel. Transfer curves could be obtained by sweeping the top-gate voltage while keeping the back-gate grounded. A maximum ON/OFF ratio of more than 10$^6$ was found when the drain-source voltage reached 500mV. An observed sub-threshold swing of 750mV/decade was comparable to that reported for MoS$_2$ field-effect transistor devices. The calculated field-effect mobility of the present device was about 1cm$^2$/Vs. The temperature-dependent transport properties were investigated using back-gated 4-terminal set-ups. For this type of device, the measured four-probe conductance at 2 to 300K and a constant back-gate voltage of 40V, showed that negligible hysteresis occurred and that non-linear behavior began to disappear above 200K. There was a possible effect of the contact resistance or Schottky barrier upon mobility estimation. When the back-gate voltage was greater than 15V, a metallic state which was associated with the metal-insulator transition appeared due to an increase in the Fermi level. The temperature-dependent field-effect mobility of this device, calculated using the field-effect mobility formula, attained its maximum value at 120K. Below this critical temperature, the mobility decreased due to scattering arising from charged impurities. This is again similar to the behavior observed in MoS$_2$ back-gate devices. An increase in temperature also led to a marked decrease in the mobility, and this was attributed to an electron–phonon scattering which predominated at high temperatures. The temperature dependence obeyed a T$^{-\gamma}$ relationship, where $\gamma$ depended on the electron–phonon coupling in the sample. For the present ReS$_2$ device, $\gamma$ was about 2.6; slightly larger than the values for MoSe$_2$ and MoS$_2$. It was suggested that additional suppression of phonon-scattering could be achieved by using a suitable substrate plus encapsulation in a high-$\kappa$ environment. The dependence of conductance upon temperature for various back-gate voltages showed that metallic behavior typically occurred at high temperatures. It seemed that the temperature-regime of metallic behavior grew with increasing back-gate voltage, as in the case of WS$_2$ and MoS$_2$. Below 120K the conductance variation weakened for all back-gate voltages. This could be attributed to the hopping of carriers through localized states, driving the system into a strongly localized regime as the hopping became predominant at lower temperatures. In the insulating regime at 70 to 250K, the temperature variation of the conductance could be modelled as an Arrhenius relationship in which the activation energy was that for the thermal activation of charge carriers, at the Fermi energy, into the conduction band. It decreased as the Fermi level moved towards the conduction band, again in agreement with data on $MoS_2$ and $WS_2$ field-effect transistor devices. At 20 to 250K, a two-dimensional variable-range hopping model, characterized by the equation, $\exp\text{-}(T_0/T)^{1/3}$, provided an excellent description of the electrical transport of $ReS_2$ nanosheets. The localization length could be determined from the expression, $(13.8/kDT_0)^{1/2}$, where D was the density-of-states. Taking the latter to be $4 \times 10^{12}/eVcm^2$ for a surface density of charge-traps at the $SiO_2$ interface, the localization-length was estimated to be about 5nm at a back-gate voltage of 15V; again consistent with $MoS_2$ and $WS_2$. The device was probed by using a focused 633nm laser beam and an illumination power of 12.5 to 1000nW. Plots of drain-source current versus drain-source voltage, with and without laser illumination, showed that there was an appreciable increase in the photocurrent at a given drain-source voltage was illuminated; indicating a marked dependence upon the laser power. This in turn was attributed to an increased number of photon-generated carriers. The time-resolved photo-response was probed by switching the laser on and off. The device underwent stable and repeatable changes under laser illumination. At a back-gate voltage of -10V and a drain-source voltage of 50mV, the device had ON/OFF currents of about 56 and 20nA, giving an ON/OFF ratio of about 2.8. The photocurrent increased with the back-gate voltage, because the high back-gate voltage could move the Fermi-level closer to the conduction band, thus making it easier for the tunnelling and thermionic currents to overcome the barrier between the channel and electrodes. The photocurrent also increased with increasing drain-source voltage, due to an increase in carrier drift-velocity and a decrease in the carrier transit time. The external quantum efficiency is an important parameter which characterises the ratio of the electrons flowing out of the device in response to impinging photons. The calculated external quantum efficiency, as a function of the incident laser power and back-gate voltage, took a maximum value of 3168%, corresponding to a back-gate voltage of 50V and an incident laser power of 25nW. This figure was comparable to that for InSe photo-detectors. For a back-gate voltage of 50V, the external quantum efficiency decreased linearly with increasing incident laser power. This decrease was attributed to the effect of trap-states caused by defects or charge impurities in the $ReS_2$ channel and adsorbents at the $ReS_2/SiO_2$ interface. An increase in the laser power could reduce the generated carriers available for extraction, due to the creation of additional traps being filled by photo-carriers, leading to saturation of the photocurrent and a drop in external quantum efficiency. The photo-response ranged up to 16.14A/W. These results showed that the performance of the present device was competitive with that of graphene, $MoSe_2$, GaS and GaSe-based photo-detectors.

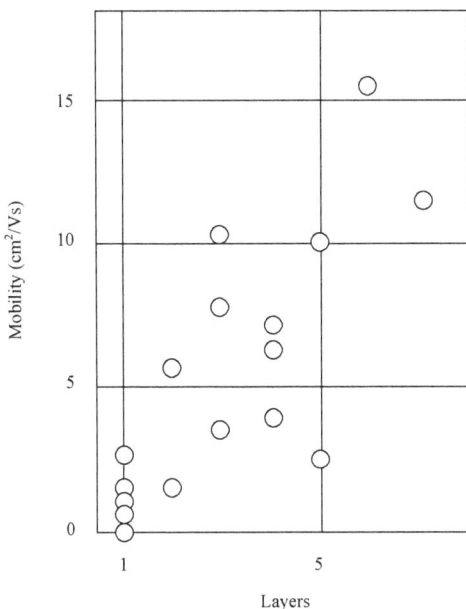

*Figure 41 Dependence of device mobility upon the number of layers*

Atomically thin rhenium disulfide flakes, with their distorted 1T structure, exhibit in-plane anisotropic properties. Monolayer and few-layer rhenium disulfide field-effect transistors have current on/off ratios of about $10^7$ and exhibit low sub-threshold swings of 100mV/decade[143]. Study of the mobility of 17 field effect transistor devices, as a function of the number of layers (figure 41) showed that the device mobility generally increases with increasing number of layers. This implies that the electrons probably travel independently through the various layers in thin disulfide flakes. The mobility of a monolayer device varies from 0.1 to $2.6cm^2/Vs$, with the highest mobility ($15.4cm^2/Vs$) being found for a six-layer device. The anisotropic ratio along two principal axes reaches 3.1. This is the greatest among the two-dimensional semiconductors. A well-performing integrated digital inverter can be constructed by combining two rhenium disulfide anisotropic field effect transistors, thus promising the potential construction of large-scale two-dimensional logic circuits. The digital inverter was made by combining two anisotropic rhenium disulfide field effect transistors along the a- and b-axes (figure 42). A quadrilateral few-layer disulfide flake having a 60° inner angle was used to prepare two field effect transistors along two axes and 15nm of $HfO_2$ was then deposited to form the top dielectric, with 30nm of gold being finally added as top-gate electrodes.

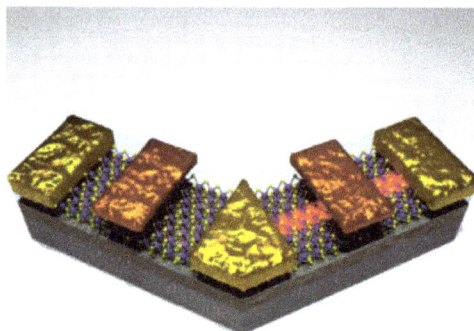

*Figure 42 Schematic structure of an inverter which combines two top-gated anisotropic ReS$_2$ field effect transistors: the left-hand FET lying along the a-axis and the right-hand FET lying along the b-axis*

Negative differential resistance devices have also attracted attention due to their folded current-voltage characteristics and consequent multiple threshold voltages. This extraordinary property makes negative differential resistance devices the promising basis of multi-valued logic applications. A negative differential resistance device has been constructed on the basis of a phosphorene/rhenium disulfide heterojunction (figure 43) involving type-III broken-gap band alignment[144]. This exhibits peak-to-valley current ratios of 4.2 and 6.9 at room temperature and 180K, respectively. The carrier transport mechanism of such a negative differential resistance device can be investigated in detail by analysing the tunnelling and diffusion currents at various temperatures. Here, the negative differential resistance behaviour was observed between 0.4 and 0.9V, with a peak-to-valley current ratio of 4.2. This was the highest value among previously reported negative differential resistance devices based upon two-dimensional materials. A ternary inverter has also been demonstrated for use in multi-valued logic applications. This study of a two-dimensional material heterojunction is a step forward toward future multi-valued logic device research.

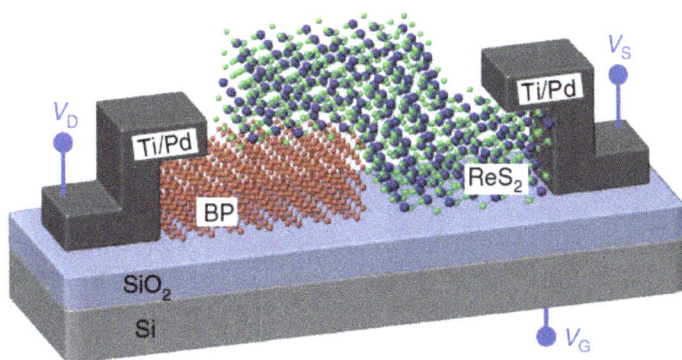

*Figure 43 Schematic illustration of a black-phosphorus/ReS$_2$ heterojunction negative differential resistance device*

### *Energy Storage*

Composites consisting of rhenium disulfide and reduced graphene oxide have been synthesized by using a simple one-pot hydrothermal method. The ReS$_2$/rGO composites exhibit an hierarchical interconnected porous morphology which is made up of nanosheets[145]. The composites were prepared by dispersing 100mg of graphene oxide into 60ml of de-ionized water and sonificating for 2h. Then 536mg of ammonium perrhenate, 417mg of hydroxylamine hydrochloride and 685mg of thiourea were dissolved in 60ml of the graphene oxide dispersion. This mixture was stirred for 2h and put into a 100ml stainless-steel autoclave which was heated to 240C and maintained there for 24h. When cooled, the black ReS$_2$/rGO powder was washed with de-ionized water and ethanol and dried. Bare ReS$_2$ was also prepared in the same way, omitting the graphene oxide. Test anodes were prepared from 80wt% of the active material, 10wt% of carbon-black and 10wt% of polyvinylidene fluoride binder. The mixture was dispersed in N-methyl-2-pyrrolidinone so as to form a slurry that was then spread onto a copper foil and dried at 120C under vacuum for 12h. A coin-type half-cell was assembled under argon, using a lithium foil as a counter-electrode, with an electrolyte of 1M LiPF$_6$ in a 1:1 v/v mixture of ethylene carbonate and dimethylcarbonate, and Celgard 2400 as a separator.

Charge/discharge tests were performed between 0.01 and 3.0V, together with cyclic voltammetry measurements at a scan-rate of 0.2mV/s between 0 and 3.0V. Electrochemical impedance spectroscopy at alternating-currents of 100kHz to 0.01Hz was also performed. When used as the anode of this lithium-ion battery, as-synthesized $ReS_2$/rGO composites offered a large initial capacity of 918mAh/g at 0.2C. The composites also exhibited much better electrochemical cycling stability and rate capability than did the bare rhenium disulfide. The marked improvement in electrochemical properties is attributed to the nanosheets and the porous structure, which permit easy electrolyte infiltration, efficient electron transfer and ionic diffusion. The electrochemical impedance spectroscopy yielded Nyquist plots which offered more detail: a high- to medium-frequency semicircle was due to the charge-transfer resistance and constant-phase capacitance of the electrode/electrolyte interface. The sloping line in the low-frequency region, representing the Warburg impedance, was related to the lithium-ion diffusion. These results indicated that the charge-transfer resistance of a $ReS_2$/rGO electrode is smaller than that of a $ReS_2$ electrode. This is attributed to the graphene component of the composite, which increases the conductivity of the electrode. The greater slope of the low-frequency line for the composite electrode also reflected the faster lithium-ion diffusion in the electrode.

Three-dimensional chrysanthemum-like microspheres comprising curly rhenium disulfide nanosheets have been produced by using a simple hydrothermal method[146]. High-resolution transmission electron micrographs indicate that the disulfide nanosheet is highly crystalline and has a thickness of only a few monolayers. When used as the anode for a lithium-ion battery, as-synthesized three-dimensional chrysanthemum-like rhenium disulfide microspheres deliver a large initial discharge capacity of 843.0mAh/g and this remains at 421.1mAh/g after 30 cycles. These values are much higher than those for commercial rhenium disulfide. The marked enhancement in electrochemical performance is attributed to the porous chrysanthemum-like microsphere structure, made up of few-layered curly disulfide nanosheets. This morphology permits easy electrolyte infiltration, efficient electron transfer and ionic diffusion. This preparation method can be extended to the preparation of other two-dimensional transition-metal chalcogenide semiconductors. It also makes rhenium disulfide a potentially key material in lithium-ion battery use.

The notoriously weak van der Waals interaction between $ReS_2$ layers, and the large interlayer spacing offers considerable possibilities for massive lithium ions to diffuse without appreciable volume expansion. In addition to these factors, this material has the most anisotropic diffusion ratio along its two principle axes as compared with any other known two-dimensional layered materials. The direction of the Re–Re atomic chain is more conductive than other crystalline orientations. Two-dimensional layered materials are always randomly oriented, with a conventional stacked geometry. Ultra-uniformly distributed vertical nanowalls have been grown onto three-dimensional graphene foam using chemical vapor deposition with Re–Re sites adjacent to the graphene in order to increase conductivity[147]. If sulfur was placed at the opening of the chemical vapor deposition quartz tube furnace, where the temperature was about 300C and argon was the carrier gas, the graphene became coated with disulfide layers which lay parallel to the graphene surface. This morphology gave inferior results as compared to the vertical arrangement. The basal planes of lateral disulfide would hinder lithium ions from diffusing efficiently as they have a high diffusion-barrier height. The vertical nanowalls expose the more active sulfur edge sites, thus improving lithium intercalation and its reversal. The vertical structure also aids the accommodation of any strain caused by lithium intercalation and de-intercalation. Three-dimensional graphene foam was chosen due to its high conductivity and high specific surface area. Use of a vertical structure also shortens the available pathways and permits the rapid diffusion of $Li^+$ ions and electrolyte ions. The disulfide content of the composite was estimated to be 77 to 87wt%, before and after incorporating it with the graphene. The above nano-walls were prepared by chemical vapor deposition, using ammonium perrhenate and a $H_2S$ flow. The nano-walls were generally less than 100nm in size and were well-organized and connected to one another. They were free-standing on the surface of the supporting graphene and formed a structure which aided the efficient infiltration of electrolyte. Dispersive X-ray spectroscopic elemental mapping showed that the carbon mapping pattern overlapped that of sulfur and rhenium, thus confirming a uniform distribution of $ReS_2$ in the carbon matrix. X-ray photo-electron spectroscopy revealed peaks, at 284.11, 41.47 and 162.13eV, which corresponded to the carbon 1s, rhenium 4f and sulfur 2p states, respectively. Compositional analysis of the peaks showed that the active material contained 9.61at%Re and 21.29at%S; close therefore to the stoichiometric values. Distinct peaks at 41.8 and 44.3eV were attributed to the rhenium $4f_{7/2}$ and $4f_{5/2}$ states, and their location indicated the presence of rhenium metal, with no trace of oxides. Peaks at 162.2 and 163.4eV were attributed to the sulfur $2p_{3/2}$ and $2p_{1/2}$ states. The numerous $ReS_2$ nanowalls were interconnected and were uniformly distributed over the graphene-foam surface, thus generating random spaces between the nanowalls, providing a higher specific surface area and promoting the rapid diffusion of lithium ions and electrons. The nanowalls had at least two lattice orientations, with interplanar distances of 0.61 and 0.2nm; corresponding to the (001) and ($1\bar{3}1$) crystal planes of $ReS_2$. Cross-sectional high-resolution transmission electron microscopy indicated the presence of some four graphene layers atop each other, with a fringe spacing of about 0.34nm. Raman spectroscopy of the composite indicated distinct peaks, at 162 and 210/cm, which

Materials Research Forum LLC
doi: http://dx.doi.org/10.21741/9781945291920

corresponded to the in-plane $E_g$ and out-of-plane $A_g$ vibrational modes, respectively, of $ReS_2$. These peaks were linked to the disulfide's low crystal symmetry and to the coupling of fundamental Raman modes with each other and to acoustic phonons. Characteristic peaks of graphene were also observed at about 1582 and 2718/cm, and their ratio indicated the presence of multilayered material. The weak peaks were indicative of highly crystalline graphene containing few defects, proving that the deposition of $ReS_2$ did not impair its quality. X-ray diffraction patterns for the pristine graphene revealed its characteristic diffraction peaks. Two distinct peaks found for the composite corresponded to $ReS_2$, indicating that it formed during chemical vapor deposition and that no side reactions occurred. The electrochemical performance of the composite as an anode for lithium-ion batteries was tested using cyclic voltammetry. The composites exhibited good stability during scanning, and a peak at about 0.8V was attributed to the formation of $Li_xReS_2$. During anodic sweeps the electrode had a strong peak at about 2.0V, due to a reversible redox reaction. The excellent cycling stability was also reflected by galvanostatic charge–discharge data. Typical discharge–charge voltage profiles at a current density of 100mA/g exhibited, during the first discharge, potential plateaux at about 0.8 and 1.0V. These were attributed, respectively, to a transformation from the semiconducting 2H phase to the metallic 1T phase and to the subsequent complete reduction of $Re^{4+}$ to rhenium nanoparticles embedded in a $Li_2S$ matrix. In the 4th and 8th discharge curves, the 0.8V potential plateau of the first discharge curve disappeared due to the insertion of $Li^+$ ions into the disulfide nanowall structure. The electrochemical impedance spectra of the composite, before and after cycling, contained a depressed semicircle in the high-frequency region and a sloping straight line in the low-frequency region. The charge-transfer resistance following cycling was lower than that before cycling, which confirming the occurrence of a phase transformation. Excellent high current-density performance was confirmed by the observation of a fast-charging process. The cycling response was such that, at current densities of 500, 800, 2000 and 5000mA/g, the composite anodes exhibited good stability, following an initial slight decay after the first few cycles. The anodes exhibited a better stability at higher current densities, and this was attributed mainly to weak interlayer interactions. At a current density of 1000mA/g, the composite electrodes still maintained a reversible capacity of 365mAh/g even after 100 cycles; implying good reversibility. After 500 cycles, the anode continued to exhibit excellent stability, with a stable capacity of over 200mAh/g. To confirm again the superiority of vertical nanowalls over lateral disulfide planes, it is instructive to note that composites of the latter type had an initial discharge capacity of 539mA/g which decreased rapidly during subsequent cycles. The difference is of course due to the greater availability of active edge-sites in the case of vertically arranged $ReS_2$

Rhenium Disulfide                                       Materials Research Forum LLC
Materials Research Foundations **40** (2018)                    doi: http://dx.doi.org/10.21741/9781945291920

as compared with a layered arrangement parallel to the graphene support. Lithium intercalation in general occurs via adsorption of the ions on the outer crystallite surface. There is then a delay due to weakening of the van der Waals forces between the top layers. At the same time there is a diffusion of absorbed $Li^+$ ions around and into the interlayer spaces, followed by the weakening of more and more layers. The lithium ions have to overcome an energy barrier, but the diffusivity can be increased by interlayer expansion and by extremely weak interlayer coupling, both of which conditions are satisfied by $ReS_2$. The considerable volume expansion of graphite which is caused by lithium intercalation is also compensated for by the very weak van der Waals forces in the rhenium disulfide. Measurements of the nanowalls following cycling showed that their microstructure was still partially intact even after 30 cycles. Three possible diffusion pathways for lithium atoms on the $ReS_2$ (001) surface were simulated using density functional theory calculations. Eight possible adsorption sites were also supposed; all located above 3 sulfur atoms. It was found that the adsorption energy of a lithium atom on a rhenium chain was the highest, so that lithium atoms tended to concentrate on the rhenium chains even though the diffusion-barrier energy of lithium atoms was lowest for non-rhenium chains. Regardless of the diffusion pathway, the diffusion-barrier energy of lithium atoms on the present surface did exceed 0.39eV. This is lower than the 0.5eV energy for pristine graphite. Due to this 0.11eV difference, the diffusion rate at 300K is increased by a factor of 50. With this lower diffusion barrier, lithium atoms could rapidly diffuse under the extremely weak van der Waals interaction. This composite material is therefore expected to find increasing use in energy storage.

## Catalysis

Monolayer transition-metal disulfides such as $ReS_2$ play an important role in catalytic processes such as the hydrofining of petroleum and a lesser role in the slurry-catalyst hydroconversion processes for up-grading heavy petroleum, coal and shale oil[148]. It is notable that this extensive, but early, review had little to recount concerning rhenium disulfide.

Two-dimensional materials have many intrinsic advantages which can be exploited in order to enhance the photocatalytic efficiency of water-splitting. *Ab initio* calculations reveal that, in monolayer and multilayer rhenium disulphide, the band-gap and band-edge positions are a very close match to the water-splitting energy levels[149]. The effective masses of the carriers are relatively low and the optical absorption coefficients are high under visible light. Due to weak interlayer coupling, such properties are independent of the layer thickness. Rhenium disulfide is a stable and efficient photocatalyst and is expected to have potential applications in the use of solar energy for water splitting.

Theoretical studies of the electronic and photocatalytic properties of single-layer rhenium disulfide under uniaxial and shear strains show that single-layer material exhibits a strongly anisotropic response to straining. It remains dynamically stable for a wide range of x-axis strains but becomes unstable above 2% y-axis compressive strain. The single-layer disulfide is predicted to be an indirect band-gap semiconductor, and there is an indirect to direct band-gap transition under during 1 to 5% x-axis tensile straining. The single-layer disulfide is not predicted to be able to catalyze the water oxidation reaction. Some 1 to 5% of y-axis tensile strain can make the single-layer disulfide capable of overall photocatalytic water splitting[150]. Single-layer material can also catalyze overall water splitting; being most efficient in aqueous solutions with a pH of 3.8.

X-ray photo-emission has been used to investigate the initial stages of rhenium disulfide oxidation, and the altered reactivity and core-level electronic structure of the defect sites produced by argon ion sputtering of monocrystalline and polycrystalline rhenium disulfide basal-plane surfaces[151]. Oxidation of the polycrystalline disulfide with $O_2$ at 100 to 300C produced a mixture of surface oxides containing rhenium in the $4^+$ and $7^+$ oxidation states (figure 44). The oxidation was made easier by the presence of low-coordination defect sites. It could reversed by re-sulfidization using $H_2S/H_2$ mixtures at 300C.

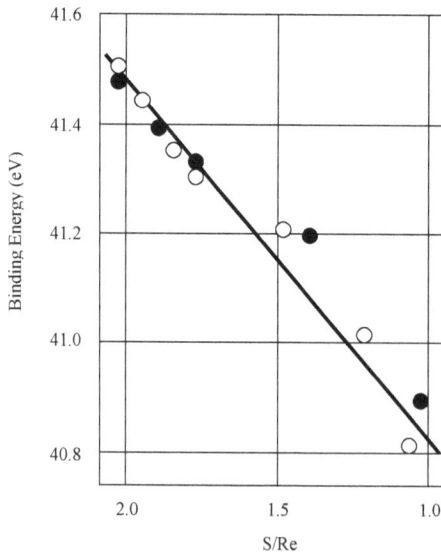

*Figure 44 Rhenium ($4f_{7/2}$) binding energy as a function of the average surface composition White circles: $ReS_2(0001)$, black circles: polycrystalline $ReS_2$*

Rhenium heptasulfide and rhenium disulfide hydrogenation catalysts have been examined for catalytic activity towards various substrates, and compared with the usual molybdenum sulfide and cobalt polysulfide catalysts[152]. Rhenium heptasulfide is the most active, and rhenium disulfide less active. The molybdenum and cobalt sulfides are much less active.

The rhenium heptasulfide hydrogenation catalyst offers consistent and easy reproducibility, the activity can be maintained indefinitely by simple storage in a closed vessel. It is stable with respect to hydrogenerative decomposition at high temperatures and is extremely resistant to poisoning. It is also insoluble in strong non-oxidizing acids, and able to saturate multiple-bond systems without any associated hydrogenolysis of carbon-sulfur bonds. On the other hand, it is not as active towards most non sulfur-containing compounds as are nickel, platinum and palladium catalysts.

Density functional theory calculations showed[153] how the structural, electronic and mechanical properties of monolayer rhenium disulfide can be tuned by hydrogenation of the surface. A stable fully-hydrogenated structure can be obtained by forming strong S-H bonds. The optimized atomic structure of $ReS_2H_2$ is quite different to that of monolayer rhenium disulfide, which is a distorted 1T phase. Phonon dispersion calculations predicted that the $Re_2$-dimerized 1T structure of $ReS_2H_2$ (termed $1TRe_2$) should be dynamically stable. Unlike the plain disulfide, the $1TRe_2$–$ReS_2H_2$ structure - formed by breaking $Re_4$ clusters into separate $Re_2$ dimers - is an indirect-gap semiconductor. Full hydrogenation enhances the flexibility of monolayer disulfide crystals but also increases the anisotropy of the elastic constants.

*Table 18 Catalytic activities for 4,6-dimethyldibenzothiophene hydro-desulfurization by rhenium catalysts*

| Catalyst | Re(wt%) | Re(at/nm$^2$) | 4,6-DMDBT(mol/gs) |
|---|---|---|---|
| Re/Al$_2$O$_3$ | 7.2 | 1.1 | $3.7 \times 10^{-8}$ |
| Re/Al$_2$O$_3$ | 11.5 | 1.8 | $8.7 \times 10^{-8}$ |
| Re/SiO$_2$ | 10 | 1.1 | $2.8 \times 10^{-7}$ |
| Re/SiO$_2$ | 19 | 2.1 | $4.8 \times 10^{-7}$ |

Rhenium disulfide-based catalysts have been prepared by using the incipient wetness impregnation method over alumina and silica supports, and evaluated for the purpose of

4,6-dimethyldibenzothiophene hydro-desulfurization in a high-pressure stirred-tank reactor. A catalyst which was prepared over silica was some 6 times more active in hydro-desulfurization than was the same catalyst when prepared over alumina[154]. It was also more active than was a $NiMo/Al_2O_3$ reference catalyst.

The unexpectedly positive $SiO_2$ support effect is attributed to the metallic nature of the supported sulfide. Opening of the S–S bond is supposed to occur during hydrogen treatment, with the resultant formation of very mobile S–H groups that are available to form molecular $H_2S$. This is confirmed by a decrease in sulfur content following hydrogen reduction (table 18).

In the case of silica-supported catalyst, the initial S/Re ratio (1.9) decreased to 1.5. This suggests the formation of highly reduced rhenium atoms. In the case of the (acidic) alumina-supported catalyst, the higher sulfur content was attributed to sulfur deposition on the carrier. Such reduction is expected to produce highly de-sulfurized photo-electron spectra-observable rhenium surface atoms (table 19) which contribute to the rhenium spectra at lower binding energies.

This is not observed. Instead, following reduction under hydrogen, there is a marked increase in the rhenium contribution at 43.0eV: 5 to 9% for the alumina-supported catalyst and 8 to 14% for the silica-supported catalyst. This corresponds to an increase in the numbers of rhenium atoms which exist in an oxysulfide environment and is attributed to the re-oxidation of de-sulfurized surface rhenium atoms by oxygen contamination.

*Table 19 Photo-electron spectra doublet binding energies, relative proportions and surface atomic ratios for $ReS_2$ supported catalysts*

| Catalyst | Re $(/nm^2)$ | Re $4f_{7/2}$(eV) | $S_{2p}[S^{2-}]$ (eV) | S/Re |
|---|---|---|---|---|
| $ReS_2/Al_2O_3$ | 1.8 | 41.7 (95%), 43.0 (5%) | 162.1, 163.3 (15%) | 3.3 |
| $ReS_2/Al_2O_3$, reduced | 1.8 | 41.9 (91%), 43.2 (9%) | 162.2, 163.3 (9%) | 2.4 |
| $ReS_2/SiO_2$ | 2.1 | 41.9 (92%), 43.0 (8%) | 162.3, 163.0 (14%) | 1.9 |
| $ReS_2/SiO_2$, reduced | 2.1 | 41.7 (85%), 43.0 (14%) | 162.2, 162.9 (9%) | 1.5 |

Thiophene hydro-desulfurization has been investigated over rhenium single crystals and polycrystalline foils and was found to be a structure-sensitive reaction over these surfaces. Rhenium monocrystals are 1 to 6 times more active than are molybdenum monocrystals; a result in agreement with studies of rhenium disulfide and molybdenum disulfide catalysts. Adsorbed carbon and sulfur overlayers decrease the activity of rhenium monocrystals, thus suggesting that the catalyst surfaces remain free of strongly bound deposits of carbon and/or sulfur[155]. The trend in hydro-desulfurization activity found for rhenium catalysts can be attributed to the difference in coordination of the top-layer rhenium atoms on monocrystalline surfaces.

An unsupported microspherical rhenium disulfide catalyst, consisting of self-assembled nano-layers, has been evaluated for the hydro-desulfurization of 3-methylthiophene and exhibited excellent catalytic activity[156]. X-ray diffraction, scanning electron microscopy, high-resolution electron microscopy, energy dispersive X-ray spectroscopy and X-ray photo-electron spectroscopy techniques showed that the rhenium disulfide layers are confined to a three-dimensional hierarchical structure with differing stackings, slab sizes and bending according to the annealing temperature (400 or 800C). The presence of a defect-rich structure in the microspheres, with short and randomly-orientated rhenium disulfide slabs, results in the exposure of additional edge sites. The latter improve the catalytic performance. Such microspherical rhenium disulfide composites, with good hydro-desulfurization performance, are promising catalysts for the desulfurization of fuel oils. The solvothermal reaction conditions also help to create exotic morphologies for the design of new rhenium disulfide catalysts.

*Table 20 Reaction rates and activation energies for $ReS_2/C$ and $CoMo/Al_2O_3$ catalysts*

| Catalyst | Temperature (C) | Reaction Rate (mol/gs) | $E_a$(kJ/mol) |
|---|---|---|---|
| $ReS_2/C$ | 280 | $3.5 \times 10^{-7}$ | 94 |
| $CoMo/Al_2O_3$ | 280 | $3.8 \times 10^{-7}$ | 75 |
| $ReS_2/C$ | 300 | $7.4 \times 10^{-7}$ | 94 |
| $CoMo/Al_2O_3$ | 300 | $6.7 \times 10^{-7}$ | 75 |
| $ReS_2/C$ | 320 | $1.48 \times 10^{-6}$ | 94 |
| $CoMo/Al_2O_3$ | 320 | $1.21 \times 10^{-6}$ | 75 |
| $ReS_2/C$ | 340 | $2.51 \times 10^{-6}$ | 94 |
| $CoMo/Al_2O_3$ | 340 | $1.91 \times 10^{-6}$ | 75 |

Ultra-small monolayer rhenium disulfide nanoplates, embedded in amorphous carbon, have been prepared by using a hydrothermal treatment involving ammonium perrhenate, thiourea, tetra-octylammonium bromide, and annealing[157]. The resultant sulfide was a low-dimensional carbon composite (ReS$_2$/C) which was tested for the purposes of hydro-desulfurization of light hydrocarbons, using 3-methylthiophene as a model molecule. This revealed the occurrence of enhanced catalytic activity as compared with that of a sulfide CoMo/Al$_2$O$_3$ catalyst (table 20). In order to understand the isomerization and hydrogenation performances of the two catalysts, the selectivity ratios of olefin/2-methyl-butane versus reaction temperature were determined (figure 45).

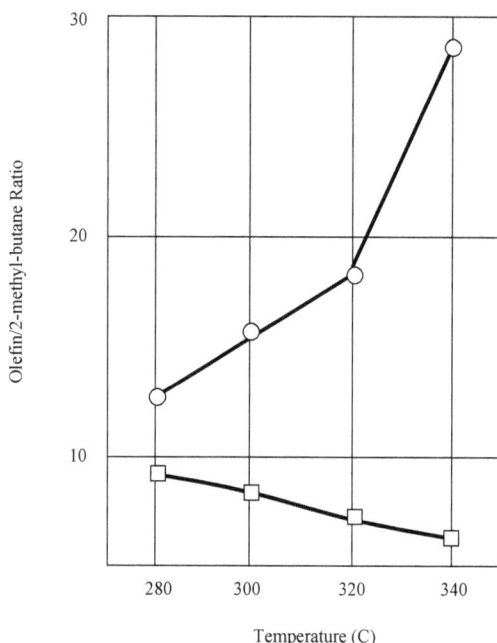

*Figure 45 Olefin/isomerization as a function of reaction temperature Circles:*
*ReS$_2$/carbon, squares: CoMo/γ-Al$_2$O$_3$*

There was a notable difference between the olefin/2-methyl-butane ratio for the two catalysts, in that there was a marked increase in the ReS$_2$/C selectivity for olefin formation as the reaction temperature increases. The opposite trend was observed for the other catalyst. The improved catalytic behaviour of the ReS$_2$/C composite was attributed to the presence of a non-stoichiometric sulfur species ($ReS_{2-x}$), to the absence of stacking along the c-axis and to the ultra-small basal planes. The latter presented a higher proportion of structural sulfur defects at the edge of the layers. This is known to be a critical factor in hydro-desulfurization catalytic processes.

The effect on activity of the activation conditions of alumina-supported rhenium disulfide catalysts for thiophene hydro-desulfurization was such that activation using $H_2S/N_2$ led to $ReS_2/\gamma-Al_2O_3$ catalysts exhibiting a greater activity than that of a NiMo catalyst[158]. Controlled-temperature reduction and X-ray photo-electron spectroscopy revealed that the higher activity of catalysts activated by $H_2S/N_2$ is associated with a higher sulfur content and was again probably related to various $ReS_x$ phases. Relevant density functional theory calculations of the reaction mechanism of thiophene hydro-desulfurization over the $ReS_2(001)$ surface under typical hydro-desulfurization reaction conditions[159] showed that the thiophene molecule takes up a so-called upright adsorption configuration, with a binding energy of 1.26eV. Two possible reaction mechanisms were considered to be direct desulfurization to butadiene or hydrogenation to 2-butene, 1-butene and butane. The results showed that hydrogen prefers to attack the thiophene carbon atoms before the first C-S bond rupture, but begins to hydrogenate the sulfur atom of thiophene following ring-opening. Pre-hydrogenation has a different effect upon the C-S bond-breaking activity. While the ring remains intact, it has little effect. When the ring is opened, suitable pre-hydrogenation can markedly decrease the energy barrier. Complete hydrogenation makes the barrier increase again due to a steric-hindrance effect. Direct de-sulfurization was proved to be kinetically unfavorable, while 2-butene was suggested to be a predominant product of the hydrogenation mechanism. Pre-adsorbed sulfur acted as a so-called ladder which aids hydrogen atoms to approach the thiophene molecule, while the thiophene sulfur atom acts as a form of intermediary for hydrogen exchange. Changing the reaction conditions by varying the $H_2$ partial pressure can affect only the rate-determining step, and not the catalytic selectivity.

Unsupported Co-Re and Ni-Re binary sulfide catalysts with atomic ratios, Me/(Me + Re), ranging from 0 to 1 were prepared by co-maceration and were characterized by using X-ray diffraction and X-ray photo-electron spectroscopy. Simultaneous hydro-desulfurization and hydrodenitrogenation reactions of a real gas oil under industrially relevant conditions revealed the occurrence of a synergetic effect of the Co or Ni sulfide upon rhenium disulfide under both hydro-desulfurization and hydrodenitrogenation conditions[160]. An activity maximum at the same atomic ratio, Me/(Me + Re) ~ 0.5, was observed for both reactions and for both the Co-Re and Ni-Re catalysts. The degree of this effect was generally greater for hydro-desulfurization than for hydrodenitrogenation and, at a low reaction temperature (598K), was greater for Ni-Re than for Co-Re catalysts; especially in the case of hydro-desulfurization. At a high temperature (648K), the trend was reversed. The Co-Re catalysts also exhibited a greater hydrodenitrogenation/ hydrodesulfurization selectivity than did the Ni-Re catalysts. Neither X-ray diffraction nor X-ray photo-electron spectroscopy revealed the presence of any Co-Re-S or Ni-Re-S type of phase. It was proposed that the synergy was associated with hydrogen activation and with spill-over from the Co or Ni sulfides to the rhenium disulfide.

The effect of the water content upon the conversion of guaiacol and phenol over a $ReS_2/SiO_2$ catalyst at 250C under a pressure of 5MPa was such that X-ray photo-electron spectroscopy revealed a loss of rhenium dispersion for catalysts containing more than 0.5ml of water[161]. The decrease in activity of the phenol conversion was related to changes in the rhenium dispersion as a result of water addition. A decrease in the guaiacol conversion was related to a loss of rhenium dispersion, together with an inhibition of the de-methylation pathway. Two forms of activated carbon (GAC, CGran Norit) were used as supports for the preparation of rhenium/carbon catalysts (0.4 rhenium atoms/nm$^2$ of support), as were Re(x)/GAC catalysts with 0.5, 0.6, 0.7 or 0.8 rhenium atoms/nm$^2$ of support. The activity difference between Re(0.4)/CGran and Re(0.4)/GAC catalysts was related to the numbers of surface oxygen functional groups on the support which were resistant to sulfidation[162]. The activity of Re(x)/GAC catalysts, as a function of rhenium content, was again related to the rhenium disulfide dispersion. The most active catalyst was that having a rhenium loading of 0.6atom/nm$^2$ of support. The selectivity, expressed in terms of the phenol/catechol ratio, did not however change with rhenium loading but was affected rather more by the acidity of the support.

A proposed proprietary rhenium disulfide nanosheet array film adsorption sensor[163] comprises an electrode base and a sensitive material. The former is composed of a gold electrode with an inter-digital electrode, and an insulating ceramic sheet substrate. The sensitive disulfide nanosheet array film is expected to be synthesized using chemical vapor deposition, and is oriented perpendicular to the electrode base so as to constitute the adsorption sensor. The latter is characterised by its simple structure, small size, low preparation costs and simplicity of synthesis. Because of the vertical orientation of the nanosheet, more gas molecules can be adsorbed on its surface thus enhancing their diffusion. Compared with existing sensors, the present sensor offers the advantages of high sensitivity, short response recovery-time and good accuracy.

*Transistors*

Transistors which are composed of exfoliated two-dimensional materials on a $SiO_2/Si$ substrate already prove effective in applications such as circuitry, memory, photo

Materials Research Forum LLC

doi: http://dx.doi.org/10.21741/9781945291920

detection, gas sensing, optical modulation, valleytronics and spintronics. The thick $SiO_2$ gate dielectric and the lack of a reliable transfer method limits however the degree of gate control. A new back-gate transistor has been fabricated[164] on an $Al_2O_3$|(indium tin oxide)|$SiO_2$|Si stack substrate and engineered with distinguishable optical identification of exfoliated two-dimensional materials. Typical two-dimensional materials, including $ReS_2$, have been used to demonstrate the increased gate controllability. Such transistors exhibit excellent electrical characteristics, including a steep sub-threshold swing (83mV/dec for $ReS_2$), high mobility (7.32cm$^2$/Vs), high ON/OFF ratio (circa $10^7$) and a reasonable working gate-bias (less than 3V). The back-gate transistor architecture was built using a 4-layer $ReS_2$ flake and the back-gate bias was applied directly to the indium tin oxide layer via the $Al_2O_3$ gate dielectric. A 200nm $SiO_2$ layer had first been grown on a heavily-doped silicon substrate by dry oxidization, and a layer of a 70nm indium tin oxide film was deposited directly onto the $SiO_2$ by sputtering (120W radio-frequency power, 550s deposition time) followed by annealing (300C, 0.25h) in order to decrease the indium tin oxide film resistance. A 25nm $Al_2O_3$ layer was then added via atomic layer deposition at 300C. The multilayer transition-metal dichalcogenides were mechanically exfoliated, using adhesive tape, onto the prepared substrate. Only those flakes with no spots or cracks, and larger than 10mm$^2$, were chosen so as to make it possible to deposit metal electrodes onto the flakes. Source and drain electrodes were designated using electron-beam lithography and chromium(10nm)/gold(70nm) was deposited (by sputtering using a radio-frequency power of 80W) in order to form the source and drain contacts. The width of the source and drain electrodes was 2mm. Oxygen plasma treatment (radio-frequency power of 30W, 30s) was used to remove any surface contaminants. The transition-metal dichalcogenide channel was directly exposed to light. The electronic characterization was carried out in air at room temperature and unipolar n-type field-effect behavior was observed plus excellent transfer characteristics, such as a cut-off current of less than 1pA, a sharp turn-on, a small gate leakage current and the above-mentioned high ON/OFF current ratio (table 21). These features, such as good mobility and remarkable sub-threshold swing were attributed to the clean fabrication process and the enhanced gate control within a small range of gate voltages which was permitted by the $Al_2O_3$/(indium tin oxide) back-gate stack. The Raman and photoluminescence spectra of 4-layer $ReS_2$ on an $Al_2O_3$|(indium tin oxide)|$SiO_2$|Si substrate revealed a broad sensitivity to visible wavelengths. The photo-response was calculated to be up to 760A/W in the depletion area and 503000A/W in the accumulation area. The ON/OFF ratio of the photocurrent was about 50 in the depletion area, 8 in the sub-threshold area and 1.2 in the accumulation area. In general, 2-dimensional material on an $Al_2O_3$|(indium tin oxide)|$SiO_2$|Si stack could absorb much more light, and generate more carriers, than that on a $SiO_2|Si$ substrate, thus leading to a marked increase in the photo-response.

*Table 21 Properties of various N-layer $ReS_2$ back-gate transistors*

| Sample | N | Mobility ($cm^2$/Vs) | Sub-Threshold Swing (mV/dec) | On/Off Ratio |
|--------|---|----------------------|------------------------------|--------------|
| 1 | 4 | 7.32 | 83 | $10^7$ |
| 2 | 4 | 6.53 | 126 | $10^6$ |
| 3 | 5 | 15.80 | 99 | $10^6$ |
| 4 | 5 | 8.45 | 145 | $10^6$ |
| 5 | 9 | 14.81 | 208 | $10^6$ |

Although rhenium disulfide films are promising candidates for opto-electronic applications because of their direct band-gap nature and optical/electrical anisotropy, the narrow spectrum and low absorption of atomically thin flakes lessen their usefulness for light-harvesting applications[1]. An efficient approach has been developed[165] for enhancing the performance of disulfide-based phototransistors by coupling CdSe-CdS-ZnS core-shell quantum dots. Under 589nm laser irradiation, the response of a rhenium disulfide phototransistor, decorated with quantum dots, could be increased more than 25-fold (up to about 654A/W). The rise and recovery times were also reduced: to 3.2 and 2.8s, respectively. The excellent opto-electronic performance was due to the coupling effect of the quantum-dot light-absorption and cross-linking 1,2-ethanedithiol ligands. Photo-excited electron-hole pairs in quantum dots could separate and transfer efficiently due to type-II band alignment and charge-exchange at the interface. Simple hybrid zero-dimensional and two-dimensional hybrid systems can thus be used for photo-detection.

A simple and efficient top-down approach has been developed, for creating a high-performance thin-film transistor and photo-detector on a thick rhenium disulfide film, by

---

[1] A word of caution is perhaps in order with regard to the related field of *mechanical-energy*-harvesting. This has its roots in long-established energy-recuperation schemes where the energy involved in the braking or the vibration-suppression of motor vehicles would be captured and fed back into the electrical storage battery. In these cases, it is obvious that the energy would otherwise have been uselessly dissipated as heat. Some designers of energy-harvesting schemes seem however to forget that Newton's third law is involved and that harvesting necessarily places a reciprocal toll on the system. In the case of recuperative braking *et cetera* that obviously does not matter. Another vehicle-related concept however, has long been that of putting ramps or rollers in roads in order to capture the supposedly 'wasted' kinetic energy. This would obviously be an entirely different matter and would amount to stealing fuel-money from the motorist. That is an extreme case but many less blatant schemes, such as using modified military footwear to charge communications equipment, would necessarily increase the fatigue of the wearer. Many designers fail to acknowledge this.

Materials Research Forum LLC
doi: http://dx.doi.org/10.21741/9781945291920

controlling its thickness[166]. Oxygen plasma treatment of the disulfide thin-film transistor led to a high ON/OFF current-ratio, high mobility, a high photo-response and rapid temporal response. Thick disulfide films were first exfoliated onto a $SiO_2$/Si substrate via mechanical cleavage using blue Nitto tape, followed by oxygen plasma-treatment of the disulfide films. The plasma treatment was performed using the parameters, 5sccm, 470mTorr and 20W; a low plasma-power being chosen in order to minimize damage to the morphology and crystal structure of the disulfide film. Differing color contrasts of the disulfide were observed, with increasing plasma treatment-time, and were related to its thickness. Photo-electron spectroscopy was used, before and after plasma treatment, in order to identify chemical changes and the formation of additional chemical bonds on the disulfide surface. Because the 2p sulfur peaks did not exhibit any appreciable change following plasma treatment, it was concluded that S-O bonds had not formed. Decreased 2p sulfur peak intensities were observed in the plasma-treated disulfide films.

A vertical resonant interlayer tunnelling field-effect transistor has been created[167] by using exfoliated few-layer rhenium disulfide flakes as electrodes and hexagonal boron nitride as the tunnel barrier. The individual 3 to 5 monolayer-thick $ReS_2$ flakes and 6 to 9 monolayer-thick hexagonal BN flakes were exfoliated onto $SiO_2$/Si substrates from bulk crystals. The flakes were picked up and layered to form $ReS_2$|hBN|$ReS_2$ stacks on the substrates by using a van der Waals layer-transfer method. The top and bottom disulfide flakes were placed with no intentional rotational alignment, and the samples were annealed in vacuum (340C, 8h, $10^{-7}$Torr) in order to remove hydrocarbon residues. Electron-beam lithography and metal deposition were used to define 40nm-thick gold on 5nm chromium contacts on each disulfide flake. The highly-doped silicon substrate served as a back-gate by which to control the carrier densities in the $ReS_2$ flakes. Two contacts were made to the bottom flake and one to the top flake. From temperature-dependent current-voltage measurements, the Schottky barrier height was estimated to be 140meV. The inter-flake current flow which was measured at higher temperatures appeared to be strongly tunnelling-limited, with a relatively small contact, and/or channel series, resistance. The location of the inter-flake current peak remained essentially fixed at 250mV, regardless of the substrate-bias and temperature, Schottky barrier properties and channel carrier concentrations and conductance. The inter-flake current appeared to become almost independent of the inter-flake voltage when below a near-zero but slightly negative value of the inter-flake voltage and above a positive value of the voltage. The fixed inter-flake currents were consistent with fixed band alignments and a fixed cathode electron concentrations resulting from electron depletion from the top and bottom flakes. For a given dielectric constant, the electric field changed due only to charge. With no top-gate, and neglecting multi-dimensional effects, the electric field above the top flake had

Materials Research Forum LLC

doi: http://dx.doi.org/10.21741/9781945291920

to vanish. With only electron charge to consider, the slope change of the electrostatic potential energy across either disulfide flake could be zero only under electron depletion or negative under electron accumulation. At negative values of the bottom-flake voltage with respect to the grounded top flake, the bottom-flake chemical potential was increased, together with the bottom-flake conduction-band edge. In the absence of a positive change in the electric displacement field across the top flake, the top-flake band edge had to remain aligned with the bottom-flake band edge and thus rose above the fixed chemical potential of the grounded top flake; consistent with electron-depletion of the top flake. Neglecting the small relative changes in the lower-flake to substrate voltage and associated electric displacement field within the thick oxide at a given substrate voltage, the charge in the bottom flake was fixed; as was required to screen out the electric displacement field between it and the substrate charge layer. With the electron concentration in the bottom flake, and the band alignment fixed, the inter-flake current was also fixed. As the bottom-flake voltage was increased, a substrate bias-dependent positive voltage was reached at which the electron charge was depleted from the bottom flake. The electric displacement field across the bottom flake became constant, the bottom conduction band edge became fixed in energy. Relative to the conduction-band edge of the top flake, the bottom-flake chemical potential fell into the band-gap of the bottom flake, and the charge then in the top flake was fixed; as required to screen out the electric displacement field between it and the substrate charge layer. With the electron concentration in the top flake, and the band alignment fixed, the inter-flake current was again fixed. According to a simple parabolic approximation, assuming an effective mass which was 0.65 times the free electron mass, the density-of-states was about 2.7 x $10^{11}$/meVcm$^2$, and the effective density-of-states was 2.3 x $10^{12}$, 3.5 x $10^{12}$ and 7.0 x $10^{12}$/cm$^2$ at 100, 150 and 300K, respectively. The combined charge density in the top and bottom flakes varied from -2.9 x $10^{12}$ to -4.3 x $10^{12}$/cm$^2$ for substrate biases ranging from 40 to 60V. The Fermi level should consequently reach a maximum of only 11 to 16 meV, respectively, in either flake at 0K; and slightly less at higher temperatures. There was no direct measurement of the ReS$_2$ mobility in this work. That the current was limited mainly by inter-flake tunnelling, even at the inter-flake current peak, led only to the requirement that the electron mobility should be greater than 0.3cm$^2$/Vs at 100 and 150K and greater than 0.2cm$^2$/Vs at 300K. This conclusion assumed a channel length-to-width ratio of 4 and an electron density of 2.9 x $10^{12}$/cm$^2$ associated with the smallest (40V) substrate-bias considered. Both conditions were easily satisfied by the collision-broadening based estimates of mobility. These mobility estimates were reasonably consistent with the measured value of 6.7cm$^2$/Vs which was found at room temperature. Due to its $\Gamma$-point conduction-band minimum, the disulfide-based system permitted

Rhenium Disulfide                                                  Materials Research Forum LLC
Materials Research Foundations **40** (2018)              doi: http://dx.doi.org/10.21741/9781945291920

resonant interlayer tunnelling, and an associated low-voltage negative differential resistance, without rotational alignment of the electrode/crystal orientations. A substantial negative differential resistance appeared to be consistent with an in-plane crystal momentum-conserving tunnelling which was considerably broadened by a scattering which was in turn consistent with low-mobility disulfide flakes.

Highly responsive phototransistors, based upon few-layer rhenium disulfide have been prepared[168]. Depending upon the back gate voltage, source drain bias and incident light intensity, the maximum photo-response could be as high as 88600A/W; two orders of magnitude higher than that of monolayer $MoS_2$. This high photo-response was attributed to increased light absorption and to gain-enhancement arising from the existence of trap-states in the few-layer disulfide flakes. It also permitted the detection of weak signals. The substrate under an hexagonal BN layer is a standard silicon wafer having a 285nm-thick oxide layer. Attention was focused on few-layer $ReS_2$ flakes with a thickness of 2.5 to 4.5nm: that is, 3 to 6 layers having an interlayer spacing of about 0.7nm. In order to reduce scattering arising from impurities in the substrate, the hexagonal BN was used rather than $SiO_2$ as the substrate. The $ReS_2$ phototransistors were first electrically characterized, in the dark, at room temperature under ambient conditions. The source-drain voltage was maintained at 1.0V. The device exhibited a very good n-type field-effect transistor behavior, with an ON/OFF ratio of up to $10^8$ and an off-state current of less than 1pA. The threshold voltage was about -30V, thus indicating either natural n-doping or a large electron concentration in the device. The n-doping was attributed to impurities or sulfur vacancies in the disulfide flakes. The field-effect mobility could be determined monitored by using the steepest slope in the transfer curves. The mobility of the few-layer devices ranged from 5 to $30cm^2/V$, and the highest mobility we found for a six-layer device. The transfer characteristics were subsequently measured under illumination. Because the direct band-gap of the disulfide ranges from 1.58eV for a monolayer to 1.5eV for the bulk, corresponding to photon wavelengths of 785 to 826.7nm, a 532nm-wavelength laser was used to generate photocurrents. The focused beam had a spot diameter of less than 1μm and, during measurement, the power of the incident radiation was kept at 20nW. The current under illumination was enhanced over the entire range, from -60 to +60V. The photo-response attained its maximum value of about 1067A/W at about 0V. It decreased monotonically with an increase to 60V or decrease to -60V. This back-gate voltage-dependent photo-response was attributed to effective tuning of the carrier density of the channel and to Schottky barriers near to the electrodes in the phototransistors. With no applied source drain bias or back-gate voltage in the dark state, due to the natural n-doping of the disulfide flakes, the Fermi level was close to the conduction-band edge. Schottky barriers formed at the interface between titanium/gold electrodes and the disulfide. Applying a source-drain bias effectively drove the photo-generated carriers between the electrodes, so that illuminating the device produced a highly efficient photocurrent collection, resulting in a large photo-response. Upon turning the device off by decreasing the back-gate voltage, the Fermi level falls into the band-gap and results in an increase in the Schottky barrier height between the electrodes and the disulfide, plus a further suppression of the generated photocurrent. Upon increasing the back-gate voltage, the device approaches a saturated state and the Fermi level moves into the conduction band. The total number of excited states is fixed under 532nm laser pumping. The approach to the saturated state indicates that most of the states in the conduction band are filled and most of the photo-generated electron–hole pairs recombine very quickly. The number of photo-generated carriers which contributes to the current is greatly reduced and this results in a decreased photo-response. The switching behavior of the channel current could be easily observed, with the response time ranging from a few seconds to tens of seconds. The current remained low when the device was kept in the dark, and increased to a much higher value under the laser illumination. Upon ceasing the illumination the current fell back to the low value. The calculated photo-response increased from 61 to 2515A/W as the bias voltage was increased from 1.0 to 5.0V. The photocurrent also increased gradually with increasing incident optical power. When the incident optical power approached 6pW, the photo-response of the few-layer $ReS_2$ phototransistor reached the above-mentioned 88600A/W while the external quantum efficiency reached $2 \times 10^7$%. Given that the photo-response of a photo-detector is equal to the product of the intrinsic response and the photoconductive gain, the high (up to 88600A/W) photo-response of the few-layer disulfide phototransistor can be attributed to an enhancement of both the quantum efficiency and the gain. The quantum efficiency is proportional to the absorbed optical power, which is in turn directly related to the film thickness. In the case of multiple-layer $ReS_2$ films, the greater thickness increases the above properties. An increased gain can be attributed to the existence of trap states arising from two main possible sources: impurities or sulfur vacancies in the disulfide flakes and surface contamination due to ambient species or preparation conditions. Under illumination, electrons which are excited from the valance band fill some of the trap-states and remain there. Due to charge conservation in the channel, holes in the valence band are continually replaced by the drain electrode when the same number of holes reaches the source electrode. Multiple holes circulate when a single photon generates an electron-hole pair. Because of the long periods during which the electrons remain in the trap states, the lifetimes of excited holes in $ReS_2$ are very long; leading to a greatly enhanced gain. The density of trap states in disulfide devices could be estimated by studying the temperature-dependence of the field-effect mobility, yielding a value of $1.96 \times 10^{13}/cm^2$. Due to the high photo-response of the present phototransistors, they are candidate materials for the detection of weak signals. A five-layer disulfide phototransistor, with the back-gate voltage fixed at -50V and the source drain bias fixed at 2.0V, yielded a photocurrent of between 1 and 7.6nA under relatively weak illumination.

Broadband polarization-sensitive photo-detectors, based upon few-layer rhenium disulfide, have been produced[169]. A transistor which was based upon few-layer material exhibited n-type behavior, with a mobility of about $40cm^2/Vs$ and an ON/OFF ratio of $10^5$. The polarization-dependence of its photo-response was attributed to its unique anisotropic in-plane crystal structure which led to optical absorption anisotropy. Linear dichroic photo-detection with a high photo-response offers a route to exploitation of the intrinsic anisotropy of two-dimensional materials and to the application of two-dimensional materials for the detection of light-polarization. Polarization-sensitive photo-detectors have also been based upon anisotropic few-layer rhenium disulfide. A transistor which was based upon few-layer disulfide exhibited n-type behavior, with a mobility of some $40cm^2/Vs$ and a photo-response of about $10^3A/W$. In order to determine the in-plane optical anisotropy, measurements were made of the polarization-dependence of the absorption spectrum of a cleaved $ReS_2$ thin flake on a quartz substrate. A large anisotropy of the absorption was observed upon varying the polarization angle from 0 to 90° with respect to the b-axis. The anisotropy was also observed from 1.55eV (800nm) to 2.76eV (450nm). This probably arose from the increased transmission of light through the ultra-thin sulfide layer. The absorption versus polarization angle could be fitted by a sinusoidal function which was identical to the band-edge absorption and Raman spectrum change for anisotropic materials. This in-plane anisotropy of the disulfide was attributed to a field-induced polarization of the lattice, leading to displacements of the lattice atoms and thus affecting the electronic states of the solid. A peak which was observed near to the band edge was related to the exciton absorption, which was also polarization-dependent; changing from 1.51 to 1.48eV upon rotating the polarization from perpendicular to parallel to the b-axis direction. This polarization-dependence of the exciton absorption could be exploited for wavelength- and polarization-sensitive luminescence applications. In order to study the photo-detector performance, field-effect transistors were prepared using atomically-thin $ReS_2$. The transistor was constructed using photolithography, following the deposition of chromium(5nm)/gold(50nm) electrodes using a high-vacuum thermal evaporator. Conduction channels were established along the b-axis direction. Data for a device based upon a flake with a thickness of about 3nm indicated that, for a low drain voltage, there was a negligible Schottky barrier at the $ReS_2$ and chromium/gold interfaces. As the gate voltage was varied from −30 to 30V, while applying a 0.1V drain voltage, the channel switched from the off-state to the on-state and there was a $10^5$ increase in the drain current. The measured ON/OFF ratio was 4 orders of magnitude greater than that of graphene and was comparable to the values reported for $MoS_2$ devices. The field-effect mobility was estimated from the linear region of current-voltage curves between 10 and 30V. The mobility was calculated to be about $18cm^2/Vs$, and this value was again comparable to the values for exfoliated $MoS_2$ layers. Further improvement of the mobility was expected to be possible by optimizing the choice of contact metals, the interface between the channel and the gate dielectric. The temperature-dependence of the mobility was studied between room temperature and 100K. Upon decreasing the temperature, the mobility increased monotonically to $40cm^2/Vs$ due to weak electron-phonon scattering at low temperatures. The temperature dependence could be described by a $T^{-\gamma}$ power-law, where $\gamma$ was about 1.1, depending upon the predominant phonon-scattering mechanism. This value was the same as the one reported for bi-layer $MoS_2$ and was smaller than that reported for monolayer $MoS_2$. Some variations in the apparent phonon damping factor could be attributed to charged-impurity scattering and homopolar phonon quenching, but the cause of the present $\gamma$-value remained unclear. The good electronic properties of $ReS_2$ make it a promising candidate for opto-electronic applications. The direct band-gap nature of the material led to a high absorption coefficient and to efficient electron–hole pair generation under photo-excitation. The photo-response of 3nm $ReS_2$ in a 2-terminal device geometry was determined by using a green 2.4eV semiconductor laser as the illumination source. The device exhibited a good response at all wavelengths. Upon increasing the illumination power, the photocurrent becomes greater. The photocurrent obeyed a simple power-law relationship with an exponent, of 0.3, which was attributed to complex processes occurring in the carrier generation, trapping and electron-hole recombination processes of the semiconductor. The process of photocurrent generation could be explained in terms of a simplified energy-band diagram. Charge-transfer via Fermi-level tuning between the metal electrode interface and $ReS_2$ channels, resulted in band-bending and the formation of Schottky-type barriers and depletion layers. Under illumination with photon energies greater than the energy-gap of the semiconductor, electron-hole pairs were excited by absorbing light and were laterally separated by the applied drain bias, thus leading to the generation of a photocurrent. The photo-response was proportional to the rate of incidence of photons. The photo-response results were several orders of magnitude better than those of graphene-based photo-detectors, comparable to the best value for $MoS_2$-based photo-detectors and 3 orders of magnitude higher than that for black phosphorus.

## Resistors

A proposed proprietary device[170] consists of a negative differential resistor which is based upon a black phosphorus/rhenium disulfide heterojunction. The device architecture consists of a silicon substrate, a silicon dioxide protective layer, a heterojunction which comprises a black phosphorus thin layer and a rhenium disulfide thin layer, a second silicon dioxide protective layer, a drain electrode and a source electrode. The silicon substrate is a grid electrode. The first silicon dioxide protective layer is grown onto the silicon substrate, and the heterojunction which comprises a black phosphorus thin layer and a rhenium disulfide thin layer, is deposited onto the first silicon dioxide protective layer. The second silicon dioxide protective layer is grown onto the heterojunction. A layer of metal is then vapor-deposited onto the surface of the second silicon dioxide protective layer in order to etch the drain electrode and the source electrode. This negative differential resistor does not require any additional doping, preparation is simpler and the semiconductors of two different materials can be combined to form the heterojunction by using the van der Waals force alone. The device possesses a high peak/valley current ratio.

## Lasers

Bulk rhenium disulfide behaves like a stack of electronically and vibrationally decoupled monolayers, thus making it easy to prepare monolayers and study their photonic properties. Due to the large layer-independent band-gap, the non-linear optical properties from visible to mid-infrared have not been fully investigated. The band structure which existed after introducing defects has been simulated using *ab initio* methods, and the results indicated that the band-gap could be reduced from 1.38 to 0.54eV by introducing defects within a suitable range[171]. Using bulk material which contained suitable defects as an example, a few-layer broadband disulfide saturable absorber was prepared via liquid-phase exfoliation. Using an as-prepared disulfide saturable absorber, passively Q-switched solid-state lasers with wavelengths of 0.64, 1.064 or 1.991µm were investigated. A femtosecond passively mode-locked laser at 1.06µm was based upon the as-prepared disulfide saturable absorber. This revealed the potential of rhenium disulfide for generating Q-switched and mode-locked pulsed lasers as well as two-dimensional opto-electronic devices having a variable band-gap.

By using the optically driving deposition method, a disulfide saturable absorber was prepared which had a modulation depth of 6.9% and a saturation fluence of 27.5µJ/cm$^2$. The nanomaterial was prepared by using the liquid exfoliation method. Disulfide powder was first dissolved in a solvent consisting of a mixture of alcohol and de-ionized water in the volume ratio of 7:3, with high-power ultrasonics being used to dissolve the powder.

Materials Research Forum LLC
doi: http://dx.doi.org/10.21741/9781945291920

Following 12h of sonification, the solution was centrifuged at 2000rpm for 20min in order to separate the nanosheets. The as-prepared dispersion was dropped onto a sapphire substrate in order to form the disulfide sample. The Raman spectrum was characterized using a 532nm excitation source, revealing peaks located at 138.1, 143.5, 450.9, 160.8, 211.8, and 234.6/cm. Because the Raman spectrum was insensitive to the number of disulfide layers, the Raman shift was suggested to be caused by the polarization of the probe laser. The sample thickness was about 5nm; corresponding to 7 layers of the disulfide. The transmittance of the product exhibited an ultra-broad absorption band beyond its band-gap. Based upon the disulfide saturable absorber, a multi-wavelength bright-dark pulse pair from a mode-locked fiber laser was observed[172]. The output spectrum had peaks which were located at 1573.5, 1591.1 and 1592.6nm, corresponding to the radio-frequency separations of 744 and 80Hz. The saturable absorbing ability of the disulfide was attributed to the formation of the bright and dark pulses. Cross-phase modulation caused by various wave-bands of bright and dark pulses supported the coexistence of bright-dark pulse pairs.

Few-layer material was prepared by using the liquid-phase method and, by using the open-aperture Z-scan method, the saturable absorption properties at 2.8μm were characterized[173] as being a saturable fluence of 22.6μJ/cm$^2$ and a modulation depth of 9.7%. A passively Q-switched 2.8μm solid-state laser was created by using the as-prepared disulfide saturable absorber. Using an absorbed pump power of 920mW, a maximum output power of 104mW was obtained with a pulse width of 324ns and a repetition rate of 126kHz. This was thought to be the first deployment of rhenium disulfide in an entirely solid-state laser.

The generation of harmonic mode-locking was demonstrated[174] in an erbium-doped laser with a microfiber-based rhenium disulfide saturable absorber. By exploiting the saturable absorption and large third-order non-linear effect of the disulfide, an harmonic mode-locked pulse with a 318.5MHz repetition rate could be obtained; corresponding to the 168th harmonic of the fundamental repetition frequency of 1.896MHz. As the pump power was gradually increased, the pulse interval remained constant while the output power increased linearly. At a pump power of 450mW, the output power was about 12mW.

Rhenium disulfide nanosheets exhibit saturable absorption at 1.55μm. By combining disulfide nanosheets with polyvinyl alcohol, a film-type saturable absorber was created[175] which demonstrated Q-switching and mode locking in erbium-doped fiber lasers. Upon tuning the pump from 45 to 120mW, the repetition rate of the Q-switched laser pulses varied from 12.6 to 19kHz and the duration changed from 23 to 5.496μs. By optimizing

Materials Research Forum LLC

doi: http://dx.doi.org/10.21741/9781945291920

the polarization state, a mode-locked state was obtained, with the emission of a train of pulses which was centered at 1558.6nm having a duration of 1.6ps and a fundamental repetition rate of 5.48MHz.

## *Medical*

Near-infrared light-mediated devices which combine medical diagnosis with clinical therapy, and which are minimally invasive but exhibit good tissue penetration, are increasing sought after; particularly in the field of cancer treatment. A sonification-assisted liquid exfoliation method has been proposed[176] for the scalable and continuous synthesis of colloidal rhenium disulfide nanosheets that could act as combined diagnostic and therapeutic agents for cancer treatment. Due to the presence of rhenium, with its high atomic number of 75 and appreciable photo-acoustic effect, PVP-capped disulfide nanosheets are potential bimodal contrast agents for computer tomographic and photo-acoustic imaging. Because of their marked near-infrared absorption and ultra-high (79.2%) photothermal conversion efficiency, disulfide nanosheets could also serve as therapeutic agents for the photothermal ablation of tumours; offering a potential elimination rate of up to 100%. The disulfide nanosheets also exhibit no measurable toxicity, according to cytotoxicity assay, serum biochemistry and histology. In a similar study, ultra-thin rhenium disulfide nanosheets were prepared using a bovine serum albumin-assisted ultrasonic exfoliation method and exhibited great biocompatibility with a high near-infrared absorbance. The high surface specific area and the presence of the bovine serum albumin led to a high loading ratio and to the modification of multifunctional molecules. A low-solubility anti-cancer drug, resveratrol, was loaded onto the ultra-thin disulfide surface in order to produce a biocompatible nanocomposite: ultra-thin disulfide-resveratrol. The targeting molecule, folic acid, was then conjoined to the bovine serum albumin molecule of ultra-thin disulfide-resveratrol to yield ultra-thin disulfide-resveratrol-folic acid. The latter exhibited a photothermal effect under 808nm laser irradiation. At a pH of 6.5, some 16.5% of the resveratrol molecules were released from the ultra-thin disulfide-resveratrol-folic acid over a period of 24h. That increased to 55.3% following six cycles of near-infrared irradiation (300s, 1Wcm$^2$). *In vitro* studies of the ultra-thin disulfide-resveratrol-folic acid showed that it exhibited low cytotoxicity plus an excellent HepG2-cell targeting ability. A greater cytotoxic effect could be achieved by optimizing the pH and temperature. *In vivo* studies of ultra-thin disulfide-resveratrol-folic acid which was intravenously injected into tumour-bearing mice showed that, after 24h, it had actively targeted and had accumulated mainly in tumour tissue. When injection was combined with three cycles of near-infrared irradiation (300s per day), the tumour was suppressed; with no relapse after 30 days.[177] An earlier study which had involved surface modification using polyethylene glycol had already had some success, with the resultant disulfide/glycol nanosheets being stable in various physiological solutions[178]. They gave good contrast in photo-acoustic imaging and X-ray computer tomographic imaging because of their strong near-infrared and X-ray absorption, respectively. Disulfide/glycol nanosheets could also be tracked via nuclear imaging following chelation-free labelling with $^{99m}Tc^{4+}$ radio-isotopic ions. Accumulation of disulfide/glycol nanosheets in tumours was observed following intravenous injection of tumour-bearing mice. Combined *in vivo* photothermal radiotherapy led to a remarkable synergistic tumour-destructive effect, with no obvious toxicity of the disulfide/glycol nanosheets being observed within 30 days.

**References**

[1]    Wilson, J.A., Yoffe, A.D., Advances in Physics, 18, 1969, 193-335. https://doi.org/10.1080/00018736900101307

[2]    Mcdonald, J.E., Cobble, J.W., Journal of Physical Chemistry, 66[5] 1962, 791-794. https://doi.org/10.1021/j100811a005

[3]    Juza, R, Biltz, W., Z., Elektrochem., 37, 1931, 498.

[4]    Ugryumova, L.E., Isakova, R.A., Snurnikova, V.A., Izv. Akad. Nauk SSSR - Neorg. Mater, 18[6] 1982, 969-972.

[5]    Zhou, J., Lin, J., Huang, X., Zhou, Y., Chen, Y., Xia, J., Wang, H., Xie, Y., Yu, H., Lei, J., Wu, D., Liu, F., Fu, Q., Zeng, Q., Hsu, C.H., Yang, C., Lu, L., Yu, T., Shen, Z., Lin, H., Yakobson, B.I., Liu, Q., Suenaga, K., Liu, G., Liu, Z., Nature, 556[7701] 2018, 355-359. https://doi.org/10.1038/s41586-018-0008-3

[6]    Kunev, D.K., Izv. Akad. Nauk. SSSR – Neorg. Mater., 14[2] 1978, 232-235.

[7]    Borowiec, J., Gillin, W.P., Willis, M.A.C., Boi, F.S., He, Y., Wen, J.Q., Wang S.L., Schulz, L., Journal of Physics – Condensed Matter, 30[5] 2018, 055702 https://doi.org/10.1088/1361-648X/aaa474

[8]    Simchi, H., Walter, T.N., Choudhury, T.H., Kirkley, L.Y., Redwing, J.M., Mohney, S.E., Journal of Materials Science, 52[17] 2017, 10127-10139. https://doi.org/10.1007/s10853-017-1228-x

[9]    Aliaga, J.A., Araya, J.F., Villarroel, R., Lozano, H., Alonso-Nu-ez, G., González, G., Journal of Coordination Chemistry, 67, 2014, 3884-3893. https://doi.org/10.1080/00958972.2014.975220

[10]   Qiu. D., Chinese Patent No. 105970297A, 2016-09-28.

[11] Gehlmann, M., Aguilera, I., Bihlmayer, G., Nemšák, S., Nagler, P., Gospodarič, P., Zamborlini, G., Eschbach, M., Feyer, V., Kronast, F., Młyńczak, E., Korn, T., Plucinski, L., Schüller, C., Blügel, S., Schneider, C.M., Nano Letters, 17[9] 2017, 5187-5192. https://doi.org/10.1021/acs.nanolett.7b00627

[12] Zhou, S., Gan, L., Wang, D., Li, H., Zhai, T., Nano Research, 11[6] 2018, 2909-2931. https://doi.org/10.1007/s12274-017-1942-3

[13] Urakami, N., Okuda, T., Hashimoto, Y., Japanese Journal of Applied Physics, 57, 2018, 02CB07 https://doi.org/10.7567/JJAP.57.02CB07

[14] Keyshar, K., Gong, Y., Ye, G., Brunetto, G., Zhou, W., Cole, D.P., Hackenberg, K., He, Y., Machado, L., Kabbani, M., Hart, A.H.C., Li, B., Galvao, D.S., George, A., Vajtai, R., Tiwary, C.S., Ajayan, P.M., Advanced Materials, 27[31] 2015, 4640-4648. https://doi.org/10.1002/adma.201501795

[15] Kim, Y., Kang, B., Choi, Y., Cho, J.H., Lee, C., 2D Materials, 4[2] 2017, 025057.

[16] Hafeez, M., Gan, L., Li, H., Ma, Y., Zhai, T., Advanced Functional Materials, 26[25] 2016, 4551-4560. https://doi.org/10.1002/adfm.201601019

[17] Xu, H., Cui, F., Wang, C., Li, X., Chinese Patent No. 105839072A, 2016-08-10.

[18] Chen, Y., Qi, F., Zheng, B., Li, P., Zhou, J., Wang, X., Zhang, W., Chinese Patent No. 105821383A, 2016-08-03.

[19] Chen, Y., Qi, F., Zheng, B., Li, P., Zhou, J., Wang, X., Zhang, W., Chinese Patent No. 106379871(A), 2017-02-08.

[20] Li, X., Cui, F., Feng, Q., Wang, G., Xu, X., Wu, J., Mao, N., Liang, X., Zhang, Z., Zhang, J., Xu, H., Nanoscale, 8[45] 2016, 18956-18962. https://doi.org/10.1039/C6NR07233J

[21] Cui, F., Wang, C., Li, X., Wang, G., Liu, K., Yang, Z., Feng, Q., Liang, X., Zhang, Z., Liu, S., Lei, Z., Liu, Z., Xu, H., Zhang, J., Advanced Materials, 28[25] 2016, 5019-5024. https://doi.org/10.1002/adma.201600722

[22] Hu, D., Xu, G., Xing, L., Yan, X., Wang, J., Zheng, J., Lu, Z., Wang, P., Pan, X., Jiao, L., Angewandte Chemie, 56[13] 2017, 3611-3615. https://doi.org/10.1002/anie.201700439

[23] Al-Dulaimi, N., Lewis, D.J., Zhong, X.L., Azad Malik, M., O'Brien, P., Journal of Materials Chemistry C, 4[12] 2016, 2312-2318. https://doi.org/10.1039/C6TC00489J

[24] Al-Dulaimi, N., Lewis, E.A., Lewis, D.J., Howell, S.K., Haigh, S.J., O'Brien, P.,

Chemical Communications, 52[50] 2016, 7878-7881.
https://doi.org/10.1039/C6CC03316D

[25]   Kang, J., Sangwan, V.K., Wood, J.D., Liu, X., Balla, I., Lam, D., Hersam, M.C.,
       Nano Letters, 16[11] 2016, 7216-7223.
       https://doi.org/10.1021/acs.nanolett.6b03584

[26]   Qi, F., Chen, Y., Zheng, B., Zhou, J., Wang, X., Li, P., Zhang, W., Materials
       Letters, 184, 2016, 324-327. https://doi.org/10.1016/j.matlet.2016.08.016

[27]   Kim, S., Yu, H.K., Yoon, S., Lee, N.S., Kim, M.H., CrystEngComm, 19[36] 2017,
       5341-5345. https://doi.org/10.1039/C7CE00926G

[28]   Zhang, T., Jiang, B., Xu, Z., Mendes, R.G., Xiao, Y., Chen, L., Fang, L.,
       Gemming, T., Chen, S., Rümmeli, M.H., Fu, L., Nature Communications, 7, 2016,
       13911. https://doi.org/10.1038/ncomms13911

[29]   Hämäläinen, J., Mattinen, M., Mizohata, K., Meinander, K., Vehkamäki, M.,
       Räisänen, J., Ritala, M., Leskelä, M., Advanced Materials, 30[24] 2018, 1703622.
       https://doi.org/10.1002/adma.201703622

[30]   Chen, Y., Qi, F., Zheng, B., Li, P., Zhou, J., Wang, X., Zhang, W., Chinese Patent
       No. 106277064A, 2017-01-04.

[31]   Qi, F., He, J., Chen, Y., Zheng, B., Li, Q., Wang, X., Yu, B., Lin, J., Zhou, J., Li,
       P., Zhang, W., Li, Y., Chemical Engineering Journal, 315, 2017, 10-17.
       https://doi.org/10.1016/j.cej.2017.01.004

[32]   Jariwala, B., Voiry, D., Jindal, A., Chalke, B.A., Bapat, R., Thamizhavel, A.,
       Chhowalla, M., Deshmukh, M., Bhattacharya, A., Chemistry of Materials, 28[10]
       2016, 3352-3359. https://doi.org/10.1021/acs.chemmater.6b00364

[33]   Yella, A., Therese, H.A., Zink, N., Panthöfer, M., Tremel, W., Chemistry of
       Materials, 20[11] 2008, 3587-3593. https://doi.org/10.1021/cm7030619

[34]   Aliaga, J.A., Araya, J.F., Lozano, H., Benavente, E., Alonso-Nu-ez, G., González,
       G., Materials Chemistry and Physics, 151, 2015, 372-377.
       https://doi.org/10.1016/j.matchemphys.2014.12.012

[35]   Aliaga, J.A., Alonso-Nú-ez, G., Zepeda, T., Araya, J.F., Rubio, P.F., Bedolla-
       Valdez, Z., Paraguay-Delgado, F., Farías, M., Fuentes, S., González, G., Journal of
       Non-Crystalline Solids, 447, 2016, 29-34.
       https://doi.org/10.1016/j.jnoncrysol.2016.05.033

[36]   Qin, J.K., Shao, W.Z., Xu, C.Y., Li, Y., Ren, D.D., Song, X.G., Zhen, L., ACS

Applied Materials and Interfaces, 9[18] 2017, 15583-15591.
https://doi.org/10.1021/acsami.7b02101

[37]    Zhang, X., Lai, Z., Liu, Z., Tan, C., Huang, Y., Li, B., Zhao, M., Xie, L., Huang, W., Zhang, H., Angewandte Chemie, 54[18] 2015, 5425-5428.
https://doi.org/10.1002/anie.201501071

[38]    Choi, J.H., Jhi, S.H., Journal of Physics - Condensed Matter, 30[10] 2018, 105403.
https://doi.org/10.1088/1361-648X/aaac95

[39]    Murray, H.H., Kelly, S.P., Chianelli, R.R., Day, C.S., Inorganic Chemistry, 33[19] 1994, 4418-4420. https://doi.org/10.1021/ic00097a037

[40]    Hou, D., Ma, Y., Du, J., Yan, J., Ji, C., Zhu, H., Journal of Physics and Chemistry of Solids, 71[11] 2010, 1571-1575. https://doi.org/10.1016/j.jpcs.2010.08.002

[41]    Hart, L., Dale, S., Hoye, S., Webb, J.L., Wolverson, D., Nano Letters, 16[2] 2016, 1381-1386. https://doi.org/10.1021/acs.nanolett.5b04838

[42]    Wen, W., Lin, J., Suenaga, K., Guo, Y., Zhu, Y., Hsu, H.P., Xie, L., Nanoscale, 9[46] 2017, 18275-18280. https://doi.org/10.1039/C7NR05289H

[43]    Dalmatova, S.A., Fedorenko, A.D., Mazalov, L.N., Asanov, I.P., Ledneva, A.Y., Tarasenko, M.S., Enyashin, A.N., Zaikovskii, V.I., Fedorov, V.E., Nanoscale, 10[21] 2018, 10232-10240. https://doi.org/10.1039/C8NR01661E

[44]    Chen, B., Wu, K., Suslu, A., Yang, S., Cai, H., Yano, A., Soignard, E., Aoki, T., March, K., Shen, Y., Tongay, S., Advanced Materials, 29[34] 2017, 1701201.
https://doi.org/10.1002/adma.201701201

[45]    Kelty, S.P., Ruppert, A.F., Chianelli, R.R., Ren, J., Whangbo, M.H., Journal of the American Chemical Society, 116[17] 1994, 7857-7863.
https://doi.org/10.1021/ja00096a048

[46]    Kelty, S.P., Ruppert, A.F., Chianelli, R.R., Ren, J.Q., Whangbo, M.H., Materials Research Society Symposium - Proceedings, 327, 1994, 59-63.
https://doi.org/10.1557/PROC-327-59

[47]    Zhuang, Y., Dai, L., Li, H., Hu, H., Liu, K., Yang, L., Pu, C., Hong, M., Liu, P., Journal of Physics D, 51[16] 2018, 165101. https://doi.org/10.1088/1361-6463/aab5a7

[48]    Enyashin, A.N., Popov, I., Seifert, G., Physica Status Solidi B, 246[1] 2009, 114-118. https://doi.org/10.1002/pssb.200844254

[49]    Sahoo, J.K., Tahir, M.N., Yella, A., Branscheid, R., Kolb, U., Tremel, W.,

Langmuir, 27[1] 2011, 385-391. https://doi.org/10.1021/la103687y

[50]  Min, Y.M., Wang, A.Q., Ren, X.M., Liu, L.Z., Wu, X.L., Applied Surface
      Science, 427, 2018, 942-948. https://doi.org/10.1016/j.apsusc.2017.09.080

[51]  Horzum, S., ÇakIr, D., Suh, J., Tongay, S., Huang, Y.S., Ho, C.H., Wu, J., Sahin,
      H., Peeters, F.M., Physical Review B, 89[15] 2014, 155433.
      https://doi.org/10.1103/PhysRevB.89.155433

[52]  Xu, K., Deng, H.X., Wang, Z., Huang, Y., Wang, F., Li, S.S., Luo, J.W., He, J.,
      Nanoscale, 7[38] 2015, 15757-15762. https://doi.org/10.1039/C5NR04625D

[53]  Mukherjee, S., Banwait, A., Grixti, S., Koratkar, N., Singh, C.V., ACS Applied
      Materials and Interfaces, 10[6] 2018, 5373-5384.
      https://doi.org/10.1021/acsami.7b13604

[54]  Gunasekera, S.M., Wolverson, D., Hart, L.S., Mucha-Kruczynski, M., Journal of
      Electronic Materials, 47[8] 2018, 4314-4320. https://doi.org/10.1007/s11664-018-
      6239-0

[55]  Zelewski, S.J., Kudrawiec, R., Scientific Reports, 7[1] 2017, 15365.
      https://doi.org/10.1038/s41598-017-15763-1

[56]  Wang, P., Wang, Y., Qu, J., Zhu, Q., Yang, W., Zhu, J., Wang, L., Zhang, W., He,
      D., Zhao, Y., Physical Review B, 97[23] 2018, 235202.
      https://doi.org/10.1103/PhysRevB.97.235202

[57]  Wang, H., Liu, E., Wang, Y., Wan, B., Ho, C.W., Miao, F., Wan, X.G., Physical
      Review B, 96[16] 2017, 165418 https://doi.org/10.1103/PhysRevB.96.165418

[58]  Schimmel, T., Friemelt, K., Lux-Steiner, M., Bucher, E., Surface and Interface
      Analysis, 23[6] 1995, 399-403. https://doi.org/10.1002/sia.740230611

[59]  Qiao, X.F., Wu, J.B., Zhou, L., Qiao, J., Shi, W., Chen, T., Zhang, X., Zhang, J.,
      Ji, W., Tan, P.H., Nanoscale, 8[15] 2016, 8324-8332.
      https://doi.org/10.1039/C6NR01569G

[60]  Nagler, P., Plechinger, G., Schüller, C., Korn, T., Physica Status Solidi R, 10[2]
      2016, 185-189. https://doi.org/10.1002/pssr.201510412

[61]  Lorchat, E., Froehlicher, G., Berciaud, S., ACS Nano, 10[2] 2016, 2752-2760.
      https://doi.org/10.1021/acsnano.5b07844

[62]  Li, T.H., Zhou, Z.H., Guo, J.H., Hu, F.R., Chinese Physics Letters, 33[4] 2016,
      046201. https://doi.org/10.1088/0256-307X/33/4/046201

[63]  Zhang, Q., Wang, W., Zhang, J., Zhu, X., Fu, L., Advanced Materials, 30[3] 2018, 1704585. https://doi.org/10.1002/adma.201704585

[64]  Ye, M., Zhang, D., Yap, Y.K., Electronics, 6[2] 2017, 43. https://doi.org/10.3390/electronics6020043

[65]  Tian, H., Tice, J., Fei, R., Tran, V., Yan, X., Yang, L., Wang, H., Nano Today, 11[6] 2016, 763-777. https://doi.org/10.1016/j.nantod.2016.10.003

[66]  Tian, H., Zhao, H., Wang, H., IEEE International Conference on Electron Devices and Solid-State Circuits, EDSSC 7785251, 2016, 234-238.

[67]  Tan, C., Lai, Z., Zhang, H., Advanced Materials, 29[37] 2017, 1701392. https://doi.org/10.1002/adma.201701392

[68]  Meng, X., Zhou, Y., Chen, K., Roberts, R.H., Wu, W., Lin, J.F., Chen, R.T., Xu, X., Wang, Y., Advanced Optical Materials, (2018) Article in press.

[69]  Wu, S., Shan, Y., Guo, J., Liu, L., Liu, X., Zhu, X., Zhang, J., Shen, J., Xiong, S., Wu, X., Journal of Physical Chemistry Letters, 8[12] 2017, 2719-2724. https://doi.org/10.1021/acs.jpclett.7b01029

[70]  Friemelt, K., Kulikova, L., Kulyuk, L., Siminel, A., Arushanov, E., Kloc, C., Bucher, E., Journal of Applied Physics, 79[12] 1996, 9268. https://doi.org/10.1063/1.362602

[71]  Yin, Y., Miao, P., Zhang, Y., Han, J., Zhang, X., Gong, Y., Gu, L., Xu, C., Yao, T., Xu, P., Wang, Y., Song, B., Jin, S., Advanced Functional Materials, 27[16] 2017, 1606694 https://doi.org/10.1002/adfm.201606694

[72]  Cui, Q., Muniz, R.A., Sipe, J.E., Zhao, H., Physical Review B, 95[16] 2017, 165406. https://doi.org/10.1103/PhysRevB.95.165406

[73]  Echeverry, J.P., Gerber, I.C., Physical Review B, 97[7] 2018, 075123. https://doi.org/10.1103/PhysRevB.97.075123

[74]  Wolverson, D., Hart, L.S., Nanoscale Research Letters, 11[1] 2016, 250. https://doi.org/10.1186/s11671-016-1459-9

[75]  Cui, F., Feng, Q., Hong, J., Wang, R., Bai, Y., Li, X., Liu, D., Zhou, Y., Liang, X., He, X., Zhang, Z., Liu, S., Lei, Z., Liu, Z., Zhai, T., Xu, H., Advanced Materials, 29[46] 2017, 1705015. https://doi.org/10.1002/adma.201705015

[76]  Gao, X., Chen, G., Li, D., Li, X., Liu, Z., Tian, J., Journal of Materials Chemistry C, 6[22] 2018, 5849-5856. https://doi.org/10.1039/C8TC01822G

[77] Shen, W., Hu, C., Tao, J., Liu, J., Fan, S., Wei, Y., An, C., Chen, J., Wu, S., Li, Y., Liu, J., Zhang, D., Sun, L., Hu, X., Nanoscale, 10[17] 2018, 8329-8337. https://doi.org/10.1039/C7NR09173G

[78] Cui, Y., Lu, F., Liu, X., Scientific Reports, 7, 2017, 40080. https://doi.org/10.1038/srep40080

[79] Cui, Q., He, J., Bellus, M.Z., Mirzokarimov, M., Hofmann, T., Chiu, H.Y., Antonik, M., He, D., Wang, Y., Zhao, H., Small, 11[41] 2015, 5565-5571. https://doi.org/10.1002/smll.201501668

[80] Arushanov, E., Bucher, E., Kloc, Ch., Kulikova, O., Kulyuk, L., Siminel, A., Proceedings of the International Semiconductor Conference, CAS, 1995, 275-278.

[81] Yang, H., Jussila, H., Autere, A., Komsa, H.P., Ye, G., Chen, X., Hasan, T., Sun, Z., ACS Photonics, 4[12] 2017, 3023-3030. https://doi.org/10.1021/acsphotonics.7b00507

[82] Aslan, O.B., Chenet, D.A., Van Der Zande, A.M., Hone, J.C., Heinz, T.F., ACS Photonics, 3[1] 2016, 96-101. https://doi.org/10.1021/acsphotonics.5b00486

[83] Friemelt, K., Lux-Steiner, M.C., Bucher, E., Journal of Applied Physics, 74[8] 1993, 5266-5268. https://doi.org/10.1063/1.354268

[84] Ho, C.H., Lee, H.W., Wu, C.C., Journal of Physics - Condensed Matter, 16[32] 2004, 5937-5944. https://doi.org/10.1088/0953-8984/16/32/026

[85] Ho, C.H., Yen, P.C., Huang, Y.S., Tiong, K.K., Journal of Physics - Condensed Matter, 13[35] 2001, 8145-8152. https://doi.org/10.1088/0953-8984/13/35/319

[86] Ho, C.H., Huang, Y.S., Tiong, K.K., Solid State Communications, 109[1] 1998, 19-22. https://doi.org/10.1016/S0038-1098(98)00519-5

[87] Hsu, H.P., Lin, K.H., Huang, Y.S., Optical Materials, 62, 2016, 433-437. https://doi.org/10.1016/j.optmat.2016.10.024

[88] Ho, C.H., Yen, P.C., Huang, Y.S., Tiong, K.K., Physical Review B, 66[24] 2002, 245207. https://doi.org/10.1103/PhysRevB.66.245207

[89] Huang, T.P., Lin, D.Y., Kao, Y.C., Wu, J.D., Huang, Y.S., Japanese Journal of Applied Physics, 50[4-2] 2011, 04DH17.

[90] Lin, D.Y., Huang, T.P., Wu, F.L., Lin, C.M., Huang, Y.S., Tiong, K.K., Solid State Phenomena, 170, 2011, 135-138. https://doi.org/10.4028/www.scientific.net/SSP.170.135

[91]    Liang, C.H., Chan, Y.H., Tiong, K.K., Huang, Y.S., Chen, Y.M., Dumcenco,
        D.O., Ho, C.H., Journal of Alloys and Compounds, 480[1] 2009, 94-96.
        https://doi.org/10.1016/j.jallcom.2008.09.175

[92]    Ho, C.H., Huang, C.E., Journal of Alloys and Compounds, 383[1-2] 2004, 74-79.
        https://doi.org/10.1016/j.jallcom.2004.04.011

[93]    Liang, C.H., Tiong, K.K., Huang, Y.S., Dumcenco, D., Ho, C.H., Journal of
        Materials Science - Materials in Electronics, 20[S1] 2009, S476-S479.
        https://doi.org/10.1007/s10854-008-9685-2

[94]    Zheng, J.Y., Lin, D.Y., Huang, Y.S., Japanese Journal of Applied Physics, 48[5]
        2009, 052302. https://doi.org/10.1143/JJAP.48.052302

[95]    Ho, C.H., Liao, P.C., Huang, Y.S., Tiong, K.K., Solid State Communications,
        103[1] 1997, 19-23. https://doi.org/10.1016/S0038-1098(97)00135-X

[96]    Huang, C.C., Kao, C.C., Lin, D.Y., Lin, C.M., Wu, F.L., Horng, R.H., Huang,
        Y.S., Japanese Journal of Applied Physics, 52[4-2] 2013, 04CH11.

[97]    Dumcenco, D.O., Huang, Y.S., Liang, C.H., Tiong, K.K., Journal of Applied
        Physics, 102[8] 2007, 083523. https://doi.org/10.1063/1.2798923

[98]    Sim, S., Lee, D., Noh, M., Cha, S., Soh, C.H., Sung, J.H., Jo, M.H., Choi, H.,
        Nature Communications, 7, 2016, 13569. https://doi.org/10.1038/ncomms13569

[99]    Sim, S., Lee, D., Noh, M., Cha, S., Soh, C.H., Sung, J.H., Choi, S., Shim, W., Jo,
        M.H., Choi, H., Optics InfoBase Conference Papers, Part F42-CLEO QELS, 2017.

[100]   Chenet, D.A., Aslan, O.B., Huang, P.Y., Fan, C., Van Der Zande, A.M., Heinz,
        T.F., Hone, J.C., Nano Letters, 15[9] 2015, 5667-5672.
        https://doi.org/10.1021/acs.nanolett.5b00910

[101]   Lin, J., Liang, L., Ling, X., Zhang, S., Mao, N., Zhang, N., Sumpter, B.G.,
        Meunier, V., Tong, L., Zhang, J., Journal of the American Chemical Society,
        137[49] 2015, 15511-15517. https://doi.org/10.1021/jacs.5b10144

[102]   Zhang, S., Mao, N., Zhang, N., Wu, J., Tong, L., Zhang, J., ACS Nano, 11[10]
        2017, 10366-10372. https://doi.org/10.1021/acsnano.7b05321

[103]   McCreary, A., Simpson, J.R., Wang, Y., Rhodes, D., Fujisawa, K., Balicas, L.,
        Dubey, M., Crespi, V.H., Terrones, M., Hight Walker, A.R., Nano Letters, 17[10]
        2017, 5897-5907. https://doi.org/10.1021/acs.nanolett.7b01463

[104]   Feng, Y., Zhou, W., Wang, Y., Zhou, J., Liu, E., Fu, Y., Ni, Z., Wu, X., Yuan, H.,
        Miao, F., Wang, B., Wan, X., Xing, D., Physical Review B, 92[5] 2015, 054110.

https://doi.org/10.1103/PhysRevB.92.054110

[105] Wen, W., Zhu, Y., Liu, X., Hsu, H.P., Fei, Z., Chen, Y., Wang, X., Zhang, M., Lin, K.H., Huang, F.S., Wang, Y.P., Huang, Y.S., Ho, C.H., Tan, P.H., Jin, C., Xie, L., Small, 13[12] 2017, 1603788. https://doi.org/10.1002/smll.201603788

[106] Ho, C.H., Liu, Z.Z., Lin, M.H., Nanotechnology, 28[23] 2017, 235203 https://doi.org/10.1088/1361-6528/aa6f51

[107] Fang, C.M., Wiegers, G.A., Haas, C., De Groot, R.A., Journal of Physics - Condensed Matter, 9[21] 1997, 4411-4424. https://doi.org/10.1088/0953-8984/9/21/008

[108] Tongay, S., Sahin, H., Ko, C., Luce, A., Fan, W., Liu, K., Zhou, J., Huang, Y.S., Ho, C.H., Yan, J., Ogletree, D.F., Aloni, S., Ji, J., Li, S., Li, J., Peeters, F.M., Wu, J., Nature Communications, 5, 2014, 3252. https://doi.org/10.1038/ncomms4252

[109] Yan, Y., Jin, C., Wang, J., Qin, T., Li, F., Wang, K., Han, Y., Gao, C., Journal of Physical Chemistry Letters, 8[15] 2017, 3648-3655. https://doi.org/10.1021/acs.jpclett.7b01031

[110] Fujita, T., Ito, Y., Tan, Y., Yamaguchi, H., Hojo, D., Hirata, A., Voiry, D., Chhowalla, M., Chen, M., Nanoscale, 6[21] 2014, 12458-12462. https://doi.org/10.1039/C4NR03740E

[111] Çakır, D., Sahin, H., Peeters, F.M., Physical Chemistry Chemical Physics, 16[31] 2014, 16771-16779. https://doi.org/10.1039/C4CP02007C

[112] Pradhan, N.R., McCreary, A., Rhodes, D., Lu, Z., Feng, S., Manousakis, E., Smirnov, D., Namburu, R., Dubey, M., Hight Walker, A.R., Terrones, H., Terrones, M., Dobrosavljevic, V., Balicas, L., Nano Letters, 15[12] 2015, 8377-8384. https://doi.org/10.1021/acs.nanolett.5b04100

[113] Gutiérrez-Lezama, I., Reddy, B.A., Ubrig, N., Morpurgo, A.F., 2D Materials, 3[4] 2016, 045016.

[114] Lin, Y.C., Komsa, H.P., Yeh, C.H., Björkman, T., Liang, Z.Y., Ho, C.H., Huang, Y.S., Chiu, P.W., Krasheninnikov, A.V., Suenaga, K., ACS Nano, 9[11] 2015, 11249-11257. https://doi.org/10.1021/acsnano.5b04851

[115] Zhong, H.X., Gao, S., Shi, J.J., Yang, L., Physical Review B, 92[11] 2015, 115438. https://doi.org/10.1103/PhysRevB.92.115438

[116] He, X., Liu, F., Hu, P., Fu, W., Wang, X., Zeng, Q., Zhao, W., Liu, Z., Small, 11[40] 2015, 5423-5429. https://doi.org/10.1002/smll.201501488

Materials Research Forum LLC
doi: http://dx.doi.org/10.21741/9781945291920

[117] Ovchinnikov, D., Gargiulo, F., Allain, A., Pasquier, D.J., Dumcenco, D., Ho, C.H., Yazyev, O.V., Kis, A., Nature Communications, 7, 2016, 12391. https://doi.org/10.1038/ncomms12391

[118] Yu, S., Zhu, H., Eshun, K., Shi, C., Zeng, M., Li, Q., Applied Physics Letters, 108[19] 2016, 191901. https://doi.org/10.1063/1.4947195

[119] Obodo, K.O., Ouma, C.N.M., Obodo, J.T., Braun, M., Physical Chemistry Chemical Physics, 19[29] 2017, 19050-19057. https://doi.org/10.1039/C7CP03455E

[120] Yen, P.C., Chen, M.J., Huang, Y.S., Ho, C.H., Tiong, K.K., Journal of Physics - Condensed Matter, 14[18] 2002, 4737-4746. https://doi.org/10.1088/0953-8984/14/18/308

[121] Ho, C.H., Optics Express, 13[1] 2005, 8-19. https://doi.org/10.1364/OPEX.13.000008

[122] Ho, C.H., Huang, Y.S., Liao, P.C., Tiong, K.K., Journal of Physics and Chemistry of Solids, 60[11] 1999, 1797-1804. https://doi.org/10.1016/S0022-3697(99)00201-2

[123] Akari, S., Friemelt, K., Glöckler, K., Lux-Steiner, M.C., Bucher, E., Dransfeld, K., Applied Physics A, 57[3] 1993, 221-223. https://doi.org/10.1007/BF00332592

[124] Schubert, B., Tributsch, H., Journal of Applied Electrochemistry, 20[5] 1990, 786-792. https://doi.org/10.1007/BF01094307

[125] Jang, H., Ryder, C.R., Wood, J.D., Hersam, M.C., Cahill, D.G., Advanced Materials, 29[35] 2017, 1700650 https://doi.org/10.1002/adma.201700650

[126] Kuc, A., Zibouche, N., Heine, T., Physical Review B, 83, 2011, 245213. https://doi.org/10.1103/PhysRevB.83.245213

[127] Webb, J.L., Hart, L.S., Wolverson, D., Chen, C., Avila, J., Asensio, M.C., Physical Review B, 96[11] 2017, 115205. https://doi.org/10.1103/PhysRevB.96.115205

[128] Biswas, D., Ganose, A.M., Yano, R., Riley, J.M., Bawden, L., Clark, O.J., Feng, J., Collins-Mcintyre, L., Sajjad, M.T., Meevasana, W., Kim, T.K., Hoesch, M., Rault, J.E., Sasagawa, T., Scanlon, D.O., King, P.D.C., Physical Review B, 96[8] 2017, 085205. https://doi.org/10.1103/PhysRevB.96.085205

[129] Aslan, O.B., Chenet, D.A., van der Zande, A.M., Hone, J.C., Heinz, T.F., ACS Photonics, 3, 2016, 96. https://doi.org/10.1021/acsphotonics.5b00486

[130] Wang, J., Yang, G., Sun, R., Yan, P., Lu, Y., Xue, J., Chen, G., Physical

Chemistry Chemical Physics, 19[39] 2017, 27052-27058.
https://doi.org/10.1039/C7CP05386J

[131] Wang, F., Yang, Z., Song, R., Optical and Quantum Electronics, 50[6] 2018, 241.
https://doi.org/10.1007/s11082-018-1510-4

[132] Park, J.Y., Joe, H.E., Yoon, H.S., Yoo, S., Kim, T., Kang, K., Min, B.K., Jun,
S.C., ACS Applied Materials and Interfaces, 9[31] 2017, 26325-26332.
https://doi.org/10.1021/acsami.7b06432

[133] Luo, M., Shen, Y.H., Song, Y.X., Japanese Journal of Applied Physics, 56[5]
2017, 055701. https://doi.org/10.7567/JJAP.56.055701

[134] Luo, M., Shen, Y.H., Yin, T.L., JETP Letters, 105[4] 2017, 255-259.
https://doi.org/10.1134/S0021364017040038

[135] Luo, M., Xu, Y.E., Optik, 158, 2018, 291-296.
https://doi.org/10.1016/j.ijleo.2017.12.134

[136] Luo, M., Xu, Y.E., Journal of Superconductivity and Novel Magnetism, 2017, in
press.

[137] Sim, S., Lee, D., Trifonov, A.V., Kim, T., Cha, S., Sung, J.H., Cho, S., Shim, W.,
Jo, M.H., Choi, H., Nature Communications, 9[1] 2018, 351.
https://doi.org/10.1038/s41467-017-02802-8

[138] Rahman, M., Davey, K., Qiao, S.Z., Advanced Functional Materials, 27[10] 2017,
1606129. https://doi.org/10.1002/adfm.201606129

[139] Qin, J.K., Shao, W.Z., Li, Y., Xu, C.Y., Ren, D.D., Song, X.G., Zhen, L., RSC
Advances, 7[39] 2017, 24188-24194. https://doi.org/10.1039/C7RA01748K

[140] Corbet, C.M., McClellan, C., Rai, A., Sonde, S.S., Tutuc, E., Banerjee, S.K., ACS
Nano, 9[1] 2015, 363-370. https://doi.org/10.1021/nn505354a

[141] Zhang, E., Jin, Y., Yuan, X., Wang, W., Zhang, C., Tang, L., Liu, S., Zhou, P., Hu,
W., Xiu, F., Advanced Functional Materials, 25[26] 2015, 4076-4082.
https://doi.org/10.1002/adfm.201500969

[142] Cho, A.J., Namgung, S.D., Kim, H., Kwon, J.Y., APL Materials, 5[7] 2017,
076101. https://doi.org/10.1063/1.4991028

[143] Liu, E., Fu, Y., Wang, Y., Feng, Y., Liu, H., Wan, X., Zhou, W., Wang, B., Shao,
L., Ho, C.H., Huang, Y.S., Cao, Z., Wang, L., Li, A., Zeng, J., Song, F., Wang, X.,
Shi, Y., Yuan, H., Hwang, H.Y., Cui, Y., Miao, F., Xing, D., Nature
Communications, 6, 2015, 6991. https://doi.org/10.1038/ncomms7991

[144] Shim, J., Oh, S., Kang, D.H., Jo, S.H., Ali, M.H., Choi, W.Y., Heo, K., Jeon, J., Lee, S., Kim, M., Song, Y.J., Park, J.H., Nature Communications, 7, 2016, 13413. https://doi.org/10.1038/ncomms13413

[145] Qi, F., Chen, Y., Zheng, B., He, J., Li, Q., Wang, X., Lin, J., Zhou, J., Yu, B., Li, P., Zhang, W., Applied Surface Science, 413, 2017, 123-128. https://doi.org/10.1016/j.apsusc.2017.03.296

[146] Qi, F., Chen, Y., Zheng, B., He, J., Li, Q., Wang, X., Yu, B., Lin, J., Zhou, J., Li, P., Zhang, W., Journal of Materials Science, 52[7] 2017, 3622-3629. https://doi.org/10.1007/s10853-016-0500-9

[147] Zhang, Q., Tan, S., Mendes, R.G., Sun, Z., Chen, Y., Kong, X., Xue, Y., Rümmeli, M.H., Wu, X.J., Chen, S., Fu, L., Advanced Materials, 28, 2016, 2616-2623. https://doi.org/10.1002/adma.201505498

[148] Chianelli, R.R., Siadati, M.H., De la Rosa, M.P., Berhault, G., Wilcoxon, J.P., Bearden, R., Abrams, B.L. Catalysis Reviews - Science and Engineering, 48[1] 2006, 1-41. https://doi.org/10.1080/01614940500439776

[149] Liu, H., Xu, B., Liu, J.M., Yin, J., Miao, F., Duan, C.G., Wan, X.G., Physical Chemistry Chemical Physics, 18[21] 2016, 14222-14227. https://doi.org/10.1039/C6CP01007E

[150] Li, Y.L., Li, Y., Tang, C., International Journal of Hydrogen Energy, 42[1] 2017, 161-167. https://doi.org/10.1016/j.ijhydene.2016.11.097

[151] Davis, S.M., Catalysis Letters, 2[1] 1989, 1-7. https://doi.org/10.1007/BF00765324

[152] Broadbent, H.S., Slaugh, L.H., Jarvis, N.L., Journal of the American Chemical Society, 76[6] 1954, 1519-1523. https://doi.org/10.1021/ja01635a016

[153] Yagmurcukardes, M., Bacaksiz, C., Senger, R.T., Sahin, H., 2D Materials, 4[3] 2017, 035013.

[154] Sepulveda, C., Garcia, R., Escalona, N., Laurenti, D., Massin, L., Vrinat, M., Catalysis Letters, 141[7] 2011, 987-995. https://doi.org/10.1007/s10562-011-0578-2

[155] Bussell, Mark E., Somorjai, Gabor A., Journal of Physical Chemistry, 93[5] 1989, 2009-2016. https://doi.org/10.1021/j100342a060

[156] Aliaga, J.A., Zepeda, T.N., Pawelec, B.N., Araya, J.F., Antúnez-García, J., Farías, M.H., Fuentes, S., Galván, D., Alonso-Nú-ez, G., González, G., Catalysis Letters,

147[5] 2017, 1243-1251. https://doi.org/10.1007/s10562-017-2024-6

[157] Aliaga, J.A., Zepeda, T., Araya, J.F., Paraguay-Delgado, F., Benavente, E., Alonso-Nú-ez, G., Fuentes, S., González, G., Catalysts, 7[12] 2017, 377 https://doi.org/10.3390/catal7120377

[158] Escalona, N., Gil Llambias, F.J., Vrinat, M., Nguyen, T.S., Laurenti, D., López Agudo, A., Catalysis Communications, 8[3] 2007, 285-288. https://doi.org/10.1016/j.catcom.2006.05.053

[159] Huang, Y., Liu, H., Chen, X., Zhou, D., Wang, C., Du, J., Zhou, T., Wang, S., Journal of Physical Chemistry C, 120[22] 2016, 12012-12021. https://doi.org/10.1021/acs.jpcc.6b02769

[160] Escalona, N., Ojeda, J., Baeza, P., García, R., Palacios, J.M., Fierro, J.L.G., Agudo, A.L., Gil-Llambías, F.J., Applied Catalysis A, 287[1] 2005, 47-53.

[161] Leiva, K., Sepúlveda, C., García, R., Fierro, J.L.G., Escalona, N., Catalysis Communications, 53, 2014, 33-37. https://doi.org/10.1016/j.catcom.2014.04.023

[162] Sepúlveda, C., García, R., Reyes, P., Ghampson, I.T., Fierro, J.L.G., Laurenti, D., Vrinat, M., Escalona, N., Applied Catalysis A, 475, 2014, 427-437.

[163] Yang, A., Chu, J., Wang, X., Lyu, P., Wang, D., Liu, D., Rong, M., Chinese Patent No. 107024516A, 2017-08-08

[164] Xu, J., Chen, L., Dai, Y.W., Cao, Q., Sun, Q.Q., Ding, S.J., Zhu, H., Zhang, D.W. Science Advances, 3[5] 2017, e1602246. https://doi.org/10.1126/sciadv.1602246

[165] Qin, J.K., Ren, D.D., Shao, W.Z., Li, Y., Miao, P., Sun, Z.Y., Hu, P., Zhen, L., Xu, C.Y., ACS Applied Materials and Interfaces, 9[45] 2017, 39456-39463. https://doi.org/10.1021/acsami.7b10349

[166] Shim, J., Oh, A., Kang, D.H., Oh, S., Jang, S.K., Jeon, J., Jeon, M.H., Kim, M., Choi, C., Lee, J., Lee, S., Yeom, G.Y., Song, Y.J., Park, J.H., Advanced Materials, 28[32] 2016, 6985-6992. https://doi.org/10.1002/adma.201601002

[167] Mohammed, O.B., Movva, H.C.P., Prasad, N., Valsaraj, A., Kang, S., Corbet, C.M., Taniguchi, T., Watanabe, K., Register, L.F., Tutuc, E., Banerjee, S.K., Journal of Applied Physics, 122[24] 2017, 245701 https://doi.org/10.1063/1.5004038

[168] Liu, E., Long, M., Zeng, J., Luo, W., Wang, Y., Pan, Y., Zhou, W., Wang, B., Hu, W., Ni, Z., You, Y., Zhang, X., Qin, S., Shi, Y., Watanabe, K., Taniguchi, T., Yuan, H., Hwang, H.Y., Cui, Y., Miao, F., Xing, D., Advanced Functional

Materials, 26[12] 2016, 1938-1944 https://doi.org/10.1002/adfm.201504408

[169] Liu, F., Zheng, S., He, X., Chaturvedi, A., He, J., Chow, W.L., Mion, T.R., Wang, X., Zhou, J., Fu, Q., Fan, H.J., Tay, B.K., Song, L., He, R.H., Kloc, C., Ajayan, P.M., Liu, Z., Advanced Functional Materials, 26[8] 2016, 1169-1177. https://doi.org/10.1002/adfm.201504546

[170] Lei, S., Shen, H., Chinese Patent No. 107039587A, 2017-08-11.

[171] Su, X., Zhang, B., Wang, Y., He, G., Li, G., Lin, N.A., Yang, K., He, J., Liu, S., Photonics Research, 6[6] 2018, 498-505. https://doi.org/10.1364/PRJ.6.000498

[172] Zhao, R., Li, G., Zhang, B., He, J., Optics Express, 26[5] 2018, 5819-5826. https://doi.org/10.1364/OE.26.005819

[173] Su, X., Nie, H., Wang, Y., Li, G., Yan, B., Zhang, B., Yang, K., He, J., Optics Letters, 42[17] 2017, 3502-3505. https://doi.org/10.1364/OL.42.003502

[174] Lu, F., Modern Physics Letters B, 31[18] 2017, 1750206. https://doi.org/10.1142/S0217984917502062

[175] Mao, D., Cui, X., Gan, X., Li, M., Zhang, W., Lu, H., Zhao, J., IEEE Journal of Selected Topics in Quantum Electronics, 24[3] 2018, 7945479. https://doi.org/10.1109/JSTQE.2017.2713641

[176] Miao, Z.H., Lv, L.-X., Li, K., Liu, P.Y., Li, Z., Yang, H., Zhao, Q., Chang, M., Zhen, L., Xu, C.Y., Small, 14[14] 2018, 1703789. https://doi.org/10.1002/smll.201703789

[177] Huang, Q., Wang, S., Zhou, J., Zhong, X., Huang, Y., RSC Advances, 8[9] 2018, 4624-4633. https://doi.org/10.1039/C7RA13454A

[178] Shen, S., Chao, Y., Dong, Z., Wang, G., Yi, X., Song, G., Yang, K., Liu, Z., Cheng, L., Advanced Functional Materials, 27[28] 2017, 1700250. https://doi.org/10.1002/adfm.201700250

# Keywords

www.ingramcontent.com/pod-product-compliance
Lightning Source LLC
Chambersburg PA
CBHW061022220326
41597CB00017BB/2251